Laboratory Methods for Sensory Evaluation of Food

ELIZABETH LARMOND
Food Research Institute
Ottawa, Ont.

RESEARCH BRANCH
CANADA DEPARTMENT OF AGRICULTURE
PUBLICATION 1637
1977

Copies of this publication may be obtained from
INFORMATION DIVISION
CANADA DEPARTMENT OF AGRICULTURE
OTTAWA
K1A 0C7

This publication replaces Publication 1284, *Methods for Sensory Evaluation of Food.*

10M-38860-12:77
Cat. No.: A73-1637/1977
ISBN 0-662-01271-2

Kromar Printing Ltd.
Contract No.: 01A05-7-38860

CONTENTS

PREFACE

Sensory evaluation is commonly used in quality assurance and new product development. It is included in the course material taught in university food science courses.

Many developments and advances have been made in the area of sensory evaluation since *Methods for Sensory Evaluation of Food* was published in 1967. No attempt was made to completely cover the subject at that time. Because of the continued demand for the original publication, especially as a teaching aid, an updated version seemed necessary. The format is similar and examples of questionnaires and analyses are included in the text.

ACKNOWLEDGMENTS

The author thanks the Literary Executor of the late Sir. Ronald A. Fisher, F.R.S. Cambridge; Dr. Frank Yates, F.R.S. Rothamstead; and Messrs. Oliver and Boyd Ltd., Edinburgh, for permission to reprint in part Table V (Statistical Chart 3) and Table XX (Statistical Chart 6) from their book *Statistical Tables for Biological, Agricultural, and Medical Research.* Thanks are also due to Iowa State University Press, Ames, Iowa, for permission to reprint Statistical Charts 4 and 7 from *Statistical Methods* by George W. Snedecor and William G. Cochran, sixth edition (c), 1967. I am also grateful to John Wiley and Sons, Inc., for permission to reprint in part Tables of Random Permutations (Statistical Chart 8) from *Experimental Designs* by William G. Cochran and Gertrude M. Cox, 1950. I acknowledge with thanks D. Reidel Publishing Company, Dordrecht, Holland, for permission to reprint Statistical Chart 5 from *Chemical Senses and Flavor* Vol. 1:121-133 by A. Kramer, G. Kahan, D. Cooper, and A. Papavasilian 1974. Lastly, I would like to thank the Institute of Food Technologists for permission to reprint Statistical Chart 1 from *Food Research 1948* by E. B. Roessler, J. Warren, and J. F. Guyman, and Statistical Chart 2 from *Food Research 1956,* Vol. 21, by E. B. Roessler, G. A. Baker, and M. A. Amerine (copyright by the Institute of Food Technologists). Finally, a special debt of gratitude to everyone who has advised and worked with me in preparing this book.

INTRODUCTION

A sensory evaluation is made by the senses of taste, smell, touch, and hearing when food is eaten. The complex sensation that results from the interaction of our senses is used to measure food quality in programs for quality control and new product development. This evaluation may be carried out by one person or by several hundred.

The first and simplest form of sensory evaluation is made at the bench by the research worker who develops the new food products. He relies on his own evaluation to determine gross differences in products. Sensory evaluation is conducted in a more formal manner by laboratory and consumer panels.

Most aspects of quality can be measured only by sensory panels, although advances are being made in the development of objective tests that measure individual quality factors. As instruments are developed to measure quality, sensory evaluation will be used to prove and standardize new objective tests.

When people are used as a measuring instrument, it is necessary to rigidly control all testing methods and conditions to overcome errors caused by psychological factors. "Error" is not synonymous with mistakes, but may include all kinds of extraneous influences. The physical and mental condition of the panelist and the influence of the testing environment affect sensory tests. For example, some people may have more flavor acuity in the morning, others in the afternoon. Even the weather can influence the disposition of panelists.

Sensory evaluation panels can be grouped into three types, highly trained experts, laboratory panels, and large consumer panels.

Highly trained experts evaluate quality, and large consumer panels are used to determine consumer reaction to a product.

Sensory tests performed by fairly large panels are valuable in predicting consumer reactions. Evaluations by experts and trained laboratory panels can be useful for control purposes, for guiding product development and improvement, and for evaluating quality. The trained panel can be particularly useful in the assessment of product changes for which there is no adequate instrumentation.

PREPARING FOR THE TEST

TESTING AREA

For sensory evaluation a special testing area is used so that distractions can be minimized and conditions can be controlled. A product could be evaluated in a corner of the laboratory or at a desk, but there is no control over interruptions and distractions, and the situation will not be constant for the next test. The testing area should be a quiet, comfortable environment. If possible, the room should be air-conditioned, and in some instances it is desirable to control humidity.

The area in which the tests are conducted is usually determined by the existing facilities. When a new facility is being constructed, the testing area should be located so that it can easily be reached by the members of the panel. The preparation area should be separate from the testing area. The panelists should not enter the preparation area, as they might gain information that would influence their judgment.

Foreign odors and odors from food preparation should be kept from the testing room. Smoking should not be permitted at any time and cosmetic odors should be avoided. The Sensory Evaluation Committee of the American Society for Testing and Materials (1968) recommends that a slight positive pressure be maintained in the testing room so that odors from surrounding areas will not enter.

TESTING SETUP

For most types of testing, except profile methods, the panelists are required to make independent judgments. In order to eliminate distraction and prevent communication among the panelists, individual booths are used. These may be simple partitions put on a table. Another inexpensive arrangement is to put the partitions on hinges so they can be collapsed when the lab bench is used for other purposes.

The usual method is to construct booths along the wall that divides the room from the preparation area (Fig. 1). This arrangement is really preferable because the product to be tested can be passed through from the preparation area to the panelists and the operators do not have to serve samples in the testing room.

Fig. 1. Taste panel area.

Some method of communication from panelist to operator must be determined. In some laboratories, a switch in each booth is connected to a light in the preparation area so the panelist can signal when he is ready for another sample. In the laboratory at the Food Research Institute (FRI), a domed hatch is used that the panelist closes toward himself to indicate that he is ready for a sample (Fig. 2). This method has provided a fairly effective means of communication and it is possible for each booth to be used three or four times during a test without the operator having to go into the testing area.

The booths may be fairly simple or very elaborate, depending on the funds available. Many laboratories have a sink and tap built into each booth for expectoration and to provide water for rinsing. Ellis (1961) has pointed out that sinks, unless properly cared for, can cause sanitation and odor problems. If a sink is used, she recommends that it be a suction type such as dentists use. A tap for rinsing water in the booth is undesirable because it is difficult to control the temperature of the water. At FRI, the water for rinsing is poured some time before the test so that it will be at room temperature.

For descriptive analysis and panel training sessions, a round table that permits discussion is required. A table that has a movable

Fig. 2. Presentation of samples through hatch.

center for passing standards back and forth is useful. Whether the panelists work at a table or in individual booths, they should be provided with comfortable chairs or stools.

The color of the booths should not influence the appearance of the product being judged. An off-white or light gray color is usually recommended.

LIGHTING

Lighting should be uniform and should not influence the appearance of the product to be tested. The type of light used should be carefully chosen if color and appearance are important factors to be judged, since many fluorescent lights distort color.

To eliminate differences in color between samples, colored lights are sometimes used. The booths in the laboratory at FRI are equipped with red lights, which are used with the room darkened. These lights have not been found to be particularly effective because differences in intensity of color are still noticeable. In some cases, the lights are dimmed so that the panelist has only enough light to see what he is doing.

Amerine et al. (1965) have pointed out that it is not known what effect colored or dim lighting has on judgment. It may have less effect on experienced judges who are accustomed to it, but inexperienced judges have expressed a dislike for testing under colored lights.

TESTING SCHEDULE

The time of day that tests are run influences results. Although this cannot be controlled if the number of tests is large, late morning and midafternoon are generally the best times for testing.

Preparation area

PREPARING SAMPLES

The type of preparation equipment located in the sample preparation area depends on the products being tested. A well-equipped kitchen is a good start, with specialized equipment being added according to need. The preparation area should have a good ventilation system for removal of cooking odors. Sufficient counter space for serving and assembling samples for presentation is an important requirement (Fig. 3).

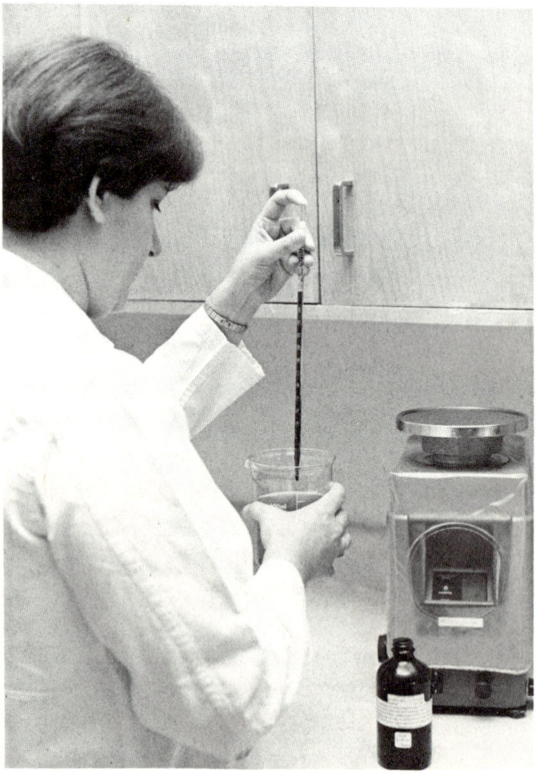

Fig. 3. Preparation of samples.

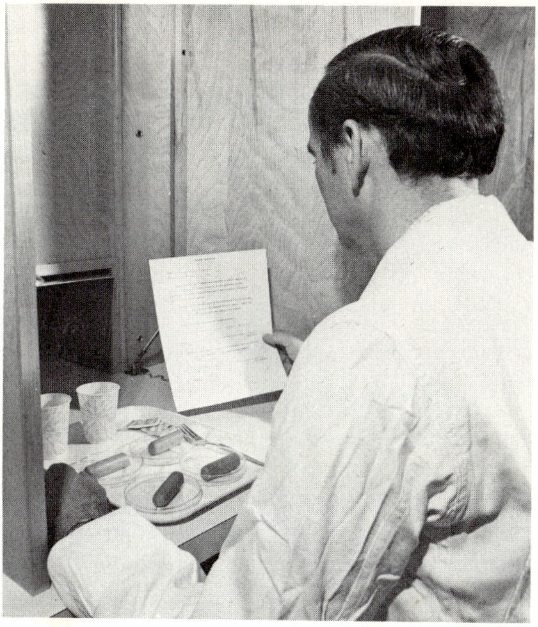

Fig. 4. Panelist with samples and questionnaire.

Preliminary testing is usually necessary to determine the method of preparation of the product. All the various factors, such as time and temperature of thawing or cooking, amount of water and size of pot for boiling, and time and speed of blending, should be predetermined and kept constant throughout tests on the product. The preparation method should not impart any foreign tastes or odors to the product.

For discrimination testing, select a preparation method that permits detection of a difference. For preference/acceptance testing, the method should be typical of normal use.

To obtain meaningful results, the samples each panelist receives must be typical of the product. It is often necessary to cut or dice and mix fresh products to get representative samples or to combine and thoroughly mix the contents of several cans or containers.

Panelists are influenced by irrelevant characteristics of the samples. Because of this, every effort should be made to make the samples from different treatments identical in all characteristics except the one being judged. It is sometimes necessary to grind, dice, chop, or puree the samples to obtain uniformity. However, Kefford and Christie (1960) found that judges prefer foods in their normal state. When conducting acceptance or preference tests, it may be better to present one sample at a time rather than grind or puree the product to mask differences.

Color differences are sometimes masked by colored or dimmed lights, as previously discussed. Some workers use colored containers such as ruby-red glasses or black-lined cups to mask color differences. Tasteless dyes have also been used to eliminate color differences.

DILUTION AND CARRIERS

Most foods are served in the way in which they are normally consumed. Some products such as hot sauces and spices must be diluted before they are tested. Spices can be mixed into a bland white sauce or a sugar syrup, but the effect of the sauce or syrup on the flavor must be considered. Several researchers have shown that hydrocolloid gels have taste-masking effects; these include Stone and Oliver (1966), Mackey and Valassi (1956), and Marshall and Vaisey (1972).

Kroll and Pilgrim (1961) found that carriers were not necessary for evaluating such products as whipped topping, cheese spread, and catsup. In fact, discrimination was better without a carrier in some cases. The use of a carrier, such as crackers for jam and frankfurters for catsup, adds to the cost and effort, and it is often difficult to select an appropriate carrier. Carriers are also a source of experimental error because the proportion of product to carrier may not be constant, or the carrier may not be of consistent quality.

The nature of a product may sometimes necessitate that it be tested with a carrier during some of the testing. Pie filling should be tested with pastry to see how the two interact, especially in terms of texture. Icing should be tested on cake not only to look at flavor combinations, but also to see how the icing handles.

SERVING TEMPERATURE

The temperature at which samples are served sometimes presents problems. For acceptance/preference tests, the samples should be served at the temperature at which they are normally eaten. As a general rule, this holds for discrimination tests also. However, it is sometimes found that panelists are able to discriminate better when the samples are slightly warmer or slightly cooler than normal. For instance, edible oils are ususally evaluated at 44°C.

Serving temperatures for many foods have been suggested by various investigators, including Caul (1957) and the Sensory Evaluation Committee of ASTM (1968). Hot foods are usually served at 60°–66°C. Ice cream should be served at –1° to 2°C and other foods at 4°–10°C. To compare results, the same temperature should be used during all tests with the product.

Once the serving temperature is determined, some method of maintaining the samples at that temperature must be found. Warming ovens with controlled temperature and humidity are available. Some laboratories use electrically heated beakers to keep samples hot. At FRI a water bath is fitted to hold beakers, and each panelist is served from the beakers as he arrives. Blocks of styrofoam can be cut out to hold containers and help maintain the temperature.

If the samples are to be held for any length of time, care must be taken that they do not dry out or change in quality during holding. When experimenting on the texture of spaghetti, for example, the spaghetti was cooled with cold water as soon as it was cooked to prevent further changes from occurring. It was cool when served to the panelists.

UTENSILS

Serving utensils should not impart any taste or odor to the product. Identical containers should be used for each sample so that no bias will be introduced from this source. Unless differences in color are being masked, it is wise to use colorless or white containers. Disposable dishes made from plastic, paper, or styrofoam are convenient when large numbers are to be served, as in consumer tests, but it must be determined beforehand that no taste is transferred to the product. For odor testing, stemmed wine glasses covered with watch glasses are often used.

Ellis (1961) recommended that a typical laboratory panel room be equipped with 96 ml *(3 oz)* ruby-red glasses, 192 ml *(6 oz)* Pyrex custard dishes, 96 ml *(3 oz)* demitasse cups, 250 to 384 ml *(8 to 12 oz)* glass snifters, 125 or 250 ml *(4 or 8 oz)* glass-stoppered bottles for odor tests, and small stainless steel forks and spoons.

QUANTITY OF SAMPLE

The amount of sample given to each panelist is often limited by the quantity of experimental material available. The Sensory Evaluation Committee of ASTM (1968) recommends that in discrimination tests, each panelist should receive at least 16 ml *(0.5 oz)* of a liquid and 28 g *(1 oz)* of a solid, and the amount should be doubled for preference tests.

The amount of sample presented should be constant throughout the testing. Panelists should receive enough sample to taste back and forth until they can make a decision. With some products such as spices, the amount the panelist tastes each time must be controlled. In work on cheese bitterness, one panel used a quinine sulfate solution as a reference and used pipettes to ensure that the same amount of solution was tasted each time.

In some tests panelists have been asked to consume fairly large amounts of a product. When conducting acceptability studies on flavored milk, the panelists were asked to drink 192 ml *(6 oz)* of the product before rating the acceptability. Such a product might be very pleasing when you drink 32 ml *(1 oz)*, but if you drink 192 ml *(6 oz)* you might find it too sweet or too satiating.

NUMBER OF SAMPLES

The number of samples that can be effectively evaluated in one session should be determined during preliminary testing. The type of product being tested and the experience of the judges must be considered when deciding on the number of samples to test in one session. Motivation is an important factor in this regard. Panelists often lose their desire to discriminate before they lose their capability.

CODING AND ORDER OF PRESENTATION

The effect of order of presentation of samples to the panelists has been investigated by many researchers. Klemmer (1968) discussed sequential effects in his paper on psychological principles of sensory evaluation. The presentation of a sample of good quality just before one of poor quality results in the rating of the second being lower than it would normally be. Similarly, if a good sample follows a poor one, it will be given a higher rating. This phenomenon is called contrast effect and is discussed later in this publication.

Because of convergence effect, which also operates when two or more samples are evaluated at the same time, a sample tends to be judged as similar to the samples it is being evaluated against, regardless of the quality. In some tests, particularly the triangle test, a positional bias has been demonstrated. When very small differences are present, there is a tendency to choose the middle sample as odd.

Because of these and other psychological and physiological effects, the order of presentation of the samples to each panelist is randomized or balanced. With a small number of samples and panelists, the order can be balanced so that every possible order occurs an equal number of times. In larger experiments, the order can be obtained from tables of random numbers. A table of random numbers is presented on page 72, Statistical Chart 8.

The code assigned to the samples should not give the panelists any hint of the identity of the treatments, and the code itself should not introduce any bias. Three-digit random numbers obtained from tables of random numbers are recommended for coding the samples. In odor tests on some products, marking pens can leave an odor on the container, so a wax pencil is used.

RINSING

The panelists are provided with an agent for oral rinsing between samples. Taste-neutral water at room temperature is preferred by many investigators. When fatty foods are being tested, warm water is a more effective rinsing agent. Crackers, apples, celery, and bread have all been used for removing flavor from the mouth. Trained cheddar cheese panelists prefer to rinse with water and then wait for 5 min before tasting the next sample. Some workers insist that the panelists rinse between each sample and others allow the panelists to rinse or not according to personal preference. Whatever the case, the panelist should consistently follow the same procedure after each sample.

The time period between samples should be constant and in some cases must be rigidly controlled. In some cases, the panelists can work at their own speed.

INFORMATION ABOUT SAMPLES

As little information as possible about the test should be given to the panelists, because this information may influence results. When a panel was told that tomato juice was made from high-quality raw ingredients, the ratings were much higher than when the same samples were presented with the information that low-quality raw products had been used (Pettit 1958). This preconceived impression is called expectation error. Panelists usually find what they expect to find. Because of this expectation error, persons who are directly involved with the experiment should not be included on the panel.

CHOOSING AND TRAINING PANELISTS

The panel is the analytical "tool" in sensory evaluation. The value of this tool depends on the objectivity, precision, and reproducibility of the judgment of the panelists. Before a panel can be used with confidence, the ability of the panelists to reproduce judgments must be determined. Interest, motivation, general attitude, and emotional state of the panelists may be responsible for inconsistent judgments.

Panelists are usually office, plant, or research staff. It should be considered a part of work routine for personnel to serve as panelists. Full cooperation of the supervisors of persons who serve as panelists is necessary. No person should be required to evaluate foods to which he objects.

The greater the number of persons on a semitrained panel, the more likely it is that individual variations will balance out. However, a small highly trained panel will give more reliable results than a large untrained panel. The minimum number of panelists should be four or five. A laboratory panel is usually composed of 10 to 20 persons with three or four replications by each judge for each treatment.

Persons who serve as panelists should be in good health and should absent themselves when suffering from conditions that might interfere with normal functions of taste and smell. Emotional factors, interest, and motivation appear to be more important than the age or sex of a panelist. It is generally recommended that panelists refrain from smoking, chewing gum, eating, or drinking for at least 30 min before testing.

The motivation of the panelist affects his response. An interested panelist is always more efficient. The panelists should be made to feel that panels are an important activity and that their contribution is important. The importance of panels can be demonstrated by running the tests in a controlled, efficient manner.

The interest of the panelists may be maintained by publishing test results. However, the panelists should feel that you are pleased to have them test and that you will be pleased no matter what the results.

Amerine et al. (1965) believe that by selecting and training the most stable and sensitive panelists, a small efficient panel can be obtained. Threshold tests are not useful in selecting panelists because sensitivity to the primary tastes may not be related to ability to detect differences in food. A more realistic approach is to select panelists on their ability to detect differences in the food to be tasted. Start with as large a group as possible and rank them according to their ability to discriminate among samples prepared, so the variations you obtain are similar to those that will be found in the actual experiment. The selected panelists should have an inherent sensitivity to the characteristic being evaluated and should be able to duplicate judgments. A new panel should be selected for each product. Persons who discriminate well on some products often do not on others.

To make objective decisions, panelists must be trained to disregard their personal preferences. The panelists must become familiar with the testing method that will be used. All panel members must know and agree upon the exact connotation of each descriptive term used. The use of physical standards during training sessions will help the panelists become more stable in their judgments. During the training period the method of handling and testing of the samples should be discussed and a common procedure agreed upon.

DESIGNING EXPERIMENTS AND CHOOSING METHODS OF ANALYZING DATA

All experiments should be planned in advance so that a simple mathematical model can be applied to the data. Most models depend on the independent nature of the data. Experimental design was introduced to ensure this independence and it often makes application of a simple mathematical model possible. Experimental designs make the test efficient and save time and material. An excellent reference on experimental designs is *Experimental Designs* by Cochran and Cox (1957).

Because of the variability in results of sensory tests, interpretation cannot be made by direct examination. The results must be tested by statistical methods. These methods compare the results actually obtained with those that would be obtained by chance alone. Results are usually expressed in degrees of significance, which is the probability that the results are caused by chance.

In this publication, the methods of analyzing the results of several sensory evaluation methods are described and several examples are given.

FACTORS INFLUENCING SENSORY MEASUREMENTS

Standard procedures for planning and conducting sensory panels have been developed in an effort to minimize or control the effect that psychological errors and physical conditions of the person or environment can have on human judgment. The necessity of standardized procedures can perhaps be emphasized by describing some of the factors that influence human judgment and illustrating the means of minimizing or eliminating them.

EXPECTATION ERROR

Any information the panelists receive about the test will influence the results. This preconceived impression is called expectation error. Panelists usually find what they expect to find. Therefore, panelists should not be given detailed information about the test and those persons who are directly involved with the experiment should not be included on the panel. The samples should be coded so that the panelists cannot identify them. The code itself should not introduce any bias. Since people generally associate "1" or "A" with "best," it is recommended that three-digit random numbers be used.

STIMULUS ERROR

In his desire to be right, the judgment of the panelist may be influenced by irrelevant characteristics of the samples. For example, when asked if there is a difference in the sweetness of two samples of peach halves, a panelist may look for help in every possible direction. He may ask himself: are the pieces of uniform size, is there a difference in color, is one firmer than the other? Because of this stimulus error all samples should be as uniform as possible. If unwanted differences occur between samples they should be masked whenever possible.

LOGICAL ERROR

Closely associated with stimulus error is logical error, which causes the panelist to assign ratings to particular characteristics because they appear to him to be logically associated with other characteristics. A slight yellow color in dehydrated potatoes, for example, may indicate oxidation to the panelist and he will logically find a different flavor in the sample. This error can be controlled by keeping the samples uniform and masking differences.

HALO EFFECT

When more than one factor in a sample is evaluated, a halo effect may be produced. The panelist often forms a general impression of a product and if asked to evaluate it for odor, texture, color, and taste at the same time, the results may be different from those when each factor is rated individually. To eliminate this effect only one characteristic should be evaluated at a time.

SUGGESTION

The response of a panelist can be influenced by the reactions of other panelists. Because of this influence, the panelists are separated from each other in individual booths. Conversation and discussion are not permitted during testing so that a suggestion from one panelist will not influence another. The testing area should be free from noise and distraction and separate from the preparation area.

MOTIVATION

The motivation of the panelist will affect his sensory perception. An interested panelist is always more efficient. The interest of the panelists can be maintained by giving them reports of their results. Trained panelists are generally more motivated than those who are not trained (Ellis 1967). The panelists should be made to feel that the panels are an important activity. This can be subtly accomplished by running the tests in a controlled, efficient manner.

CONTRAST EFFECT

The presentation of a sample of good quality just before one of poor quality causes the panelist to rate the second sample lower than it would normally be rated. Similarly, if a good sample follows a poor one it will be given a rating higher than expected. This phenomenon is called contrast effect. The order of presentation of the samples should be randomized for each panelist so that contrast effect will be equalized.

POSITIONAL BIAS

In some tests, particularly the triangle test, a positional bias has been demonstrated. When very small differences are found, panelists have a tendency to choose the middle sample as being different. This phenomenon can also be eliminated by random presentation.

METHODS FOR SENSORY TESTING

Several different sensory evaluation methods have been developed. The experimenter should be thoroughly familiar with the advantages and disadvantages of each method. The most practical and efficient method should be selected for each situation. No one method can be used universally. The experimenter must precisely define the purpose of the test and the information he wants to acquire.

There are three fundamental types of sensory tests: preference/acceptance tests, discriminatory tests, and descriptive tests. Preference/acceptance tests are affective tests based on a measure of preference or a measure from which relative preference can be determined. The personal feeling of a panelist toward the product directs his response. Discriminatory tests are used to determine whether a difference exists between samples. The panelist does not allow his personal likes and dislikes to influence his response. Laboratory difference panels can be used to determine if there is a difference among samples. Descriptive tests are used to determine the nature and intensity of the differences.

In this publication several commonly used experimental methods are described with examples of the questionnaires, their application, and statistical analyses.

DIFFERENCE TESTS

The tests to determine a difference between samples include the triangle test, the simple paired comparisons test, the Scheffé paired comparisons test, the duo-trio test, the multiple comparisons test, ranking, scoring, and ratio-scaling. Examples of these tests and sample questionnaires follow.

Triangle test

The panelist receives three coded samples. He is told that two of the samples are the same and one is different and he is asked to identify the odd sample. This method is very useful in quality control work to ensure that samples from different production lots are the same. It is also used to determine if ingredient substitution or some other change in manufacturing results in a detectable difference in the product. The triangle test is often used for selecting panelists.

Analysis of the results of triangle tests is based on the probability that if there is no detectable difference, the odd sample will be selected by chance one-third of the time.

Tables for rapid analysis of triangle test data were prepared by Roessler et al. (1948) and are reprinted on page 63. As the number of judgments increases, the percentage of correct responses required for significance decreases. For this reason, when only a small number of panelists are available, they should perform the triangle test more than once in order to obtain more judgments.

The results of a triangle test indicate whether or not there is a detectable difference between two samples. Higher levels of significance do not indicate that the difference is greater but that there is less probability of saying there is a difference when in fact there is none.

A sample of the questionnaire and an example of the triangle test follow.

QUESTIONNAIRE FOR TRIANGLE TEST

NAME____________________DATE____________________

PRODUCT____

Two of these three samples are identical, the third is different.

1. Taste the samples in the order indicated and identify the odd sample.

Code	Check odd sample
314	__________________
628	__________________
542	__________________

2. Indicate the degree of difference between the duplicate samples and the odd sample.

   Slight __________
   Moderate __________
   Much __________
   Extreme __________

3. Acceptability:

   Odd sample more acceptable __________
   Duplicates more acceptable __________

4. Comments:

EXAMPLE. A triangle test was used to determine if there was a detectable difference between fish–potato flakes processed under two different sets of conditions.

The reconstituted samples were served in coded dishes to 12 judges. Each judge received three coded samples: six judges tested two samples from treatment A and one from treatment B, and the other six judges received one sample from treatment A and two from treatment B. Because of the balanced presentation, it was necessary to assign two code numbers to each treatment.

Treatment	Code	Number of samples required
A	314	6
	542	12
B	628	12
	149	6

The order of the three samples was randomized for each judge and the order in which each judge should taste the samples was indicated by putting the code numbers in the appropriate order on the score sheet.

The odd sample was correctly identified by nine judges. According to Statistical Chart 1 in the Appendix, page 63, nine correct judgments out of 12 in a triangle test indicate a significant difference at the 1% level. The conclusion is that there is a detectable difference between the samples. If the number of correct judgments had been less than eight, the conclusion would have been that no detectable difference existed between the samples.

The degree of difference indicated by the nine judges who correctly identified the odd sample was:

	Number of judges	Value
Slight	3	× (1) = 3
Moderate	3	× (2) = 6
Much	2	× (3) = 6
Extreme	1	× (4) = 4
Total		19
Mean		2.1

The average difference was 2.1 (moderate).

The next part of the triangle test was to choose the more acceptable sample. Of the nine judges who correctly identified the odd sample, six found the samples from Treatment A more acceptable. According to Statistical Chart 2, for a two-sample test (there are only two choices at this point), six is below the number required for significance at the 5% level. However, the same degree of importance cannot be attached to the results of this secondary question as to the primary question. When it is determined that a difference exists, another test should be conducted to determine which sample is superior.

Simple paired comparisons test

A pair of coded samples is presented for comparison on the basis of some specified characteristic such as sweetness. This method has applications similar to the triangle test. Fewer samples are required and there is less tasting, but the statistical efficiency is not as great. The probability of a panelist selecting a sample by chance is 50%. Roessler et al. (1956) published tables for the rapid analysis of the results of paired comparisons tests. Two tables for paired comparisons data are presented. One table is used for directional difference tests, or one-tailed tests, when only one answer is correct. The other table is used in preference tests, or two-tailed tests, when either response can be correct. Paired comparisons tests give no indication of the size of the difference between the two samples but determine whether there is a detectable difference or not.

A sample questionnaire for a simple paired comparisons test follows.

QUESTIONNAIRE FOR
SIMPLE
PAIRED COMPARISONS TEST

NAME____________________ DATE____________________

PRODUCT____

Evaluate the sweetness of these two samples of canned peaches. Taste the sample on the left first. Indicate which sample is sweeter.

581	716
___	___

Comments:

EXAMPLE. To determine if there was a difference in sweetness between peaches canned in liquid sucrose at 45 Brix or in a 52% invert syrup at 45 Brix, a paired comparisons test was used.

Peaches from each treatment were served in coded dishes to 20 judges. Half the judges were asked to taste one sample first, the others to taste it second. Twelve judges chose the 52% invert syrup sample as sweeter. According to Statistical Chart 2 in the Appendix on page 64, in a two-sample difference test one sample must be selected 15 times out of 20 judgments to be significantly different at the 5% level of significance. The conclusion is that there is no detectable difference in sweetness between the two treatments.

Multiple paired comparisons test

When there are more than two samples to be evaluated, each must be compared with every other sample. The number of pairs is determined by the formula $\frac{1}{2}\, n(n-1)$, where n = the number of samples or treatments. This test is called a multiple paired comparisons test.

The results of multiple paired comparisons cannot be analyzed by using the same statistical tables as simple paired comparisons. At Agriculture Canada, computer programs based on the Bradley-Terry (1952) model have been developed (Dietrich 1968). The program gives maximum likelihood estimates with necessary information for performing Tukey's Test on the $\log_e$ of the estimates.

The maximum likelihood estimates of the treatments always total 1.00. These values indicate how the treatments stand in relation to one another but they give no indication of the quality of the samples.

Scheffé paired comparisons test

Scheffé (1952) modified the paired comparisons test to ask the panelists to indicate the size of the difference detected.

The results are analyzed by an analysis of variance. An average value (parameter) for each treatment is calculated. The values of the parameters are relative and the sum of the parameters for all the treatments must total 0.

An example of the questionnaire and analysis of results follow.

QUESTIONNAIRE FOR
SCHEFFÉ PAIRED COMPARISONS TEST

NAME______________________________DATE__________________________

Examine these two samples of barbecued chicken for juiciness.

1. Indicate the degree of difference in juiciness between the two samples by checking one of the following statements.

846 is extremely more juicy than 165	____
846 is much more juicy than 165	____
846 is slightly more juicy than 165	____
no difference	____
165 is slightly more juicy than 846	____
165 is much more juicy than 846	____
165 is extremely more juicy than 846	____

2. Rate the juiciness of each sample.

846		165	
____	very dry	____	very dry
____	moderately dry	____	moderately dry
____	slightly dry	____	slightly dry
____	slightly juicy	____	slightly juicy
____	moderately juicy	____	moderately juicy
____	very juicy	____	very juicy

Comments:

EXAMPLE. The Scheffé paired comparisons test was used to study the effect of four different holding temperatures on the juiciness of barbecued chicken.

Samples from each of the four treatments were compared with every other treatment, making a total of six comparisons. Each pair was presented to six judges for evaluation according to the questionnaire. The experimental design requires that half the judges taste one sample of a pair first and that the others taste the second sample first. The order of the six pairs was randomized for each judge.

The descriptive terms of the paired comparison question are assigned numerical values of +3, +2, +1, 0, −1, −2, −3.

If samples A and B are coded 846 and 165 respectively, three judges are asked to test 846 first and the score sheet is made up:

846 is extremely more juicy than 165 (+3)
846 is much more juicy than 165 (+2)
846 is slightly more juicy than 165 (+1)
no difference (0)
165 is slightly more juicy than 846 (−1)
165 is much more juicy than 846 (−2)
165 is extremely more juicy than 846 (−3)

The values assigned by the experimenter are shown on the right hand side.

The other three judges are asked to test 165 first and the score sheet is made up:

165 is extremely more juicy than 846 (+3)
165 is much more juicy than 846 (+2)
165 is slightly more juicy than 846 (+1)
no difference (0)
846 is slightly more juicy than 165 (−1)
846 is much more juicy than 165 (−2)
846 is extremely more juicy than 165 (−3)

The corresponding numerical values are shown on the right.

The following table shows the ratings assigned to each pair by the judges. The order in which the two samples of a pair are presented is taken into account.

Order of presentation	−3	−2	−1	0	1	2	3	Total score	Mean	Average preference
A–B		1	2					−4	−1.33	−1.83
B–A						2	1	7	2.33	
A–C			2		1			−1	−0.33	−0.83
C–A				1		2		4	1.33	
A–D					3			3	1.00	0.17
D–A			1		1	1		2	0.67	
B–C					2	1		4	1.33	1.67
C–B	1	1	1					−6	−2.00	
B–D				1		2		4	1.33	0.83
D–B			1	2				−1	−0.33	
C–D		1		1	1			−1	−0.33	0.34
D–C		2			1			−3	−1.00	
Total	1	5	7	5	9	8	1			

The total score for each row is obtained by multiplying each value by its frequency and adding these values.

The mean for each row = total score/number of judges

$$= -4/3$$
$$= -1.33$$

The average preference = ½ × the mean score for A over B by the first three judges (−1.33) plus the mean score for A over B by the second three judges (−2.33). The value 2.33 is the mean score of B over A by the second three judges. To determine the mean score of A over B, use the opposite score (−2.33).

$$= \tfrac{1}{2}(-1.33 - 2.33)$$
$$= \tfrac{1}{2}(-3.66)$$
$$= -1.83$$

The data from the table were used to perform an analysis of variance according to the method of Scheffé (1952).

Main effects of treatments ($\hat{\alpha}$) were calculated by obtaining the mean of the average preference of each sample over every other sample.

$\hat{\alpha}_A$ = (average preference of A over B + average preference of A over C + average preference of A over D)/number of treatments

$$\hat{\alpha}_A = [-1.83 + (-0.83) + 0.17]/4$$
$$= -2.49/4$$
$$= -0.6225$$
$$\hat{\alpha}_B = (1.83 + 1.67 + 0.83)/4$$
$$= 4.33/4$$
$$= 1.0825$$
$$\hat{\alpha}_C = [0.83 + (-1.67) + 0.34]/4$$
$$= -0.50/4$$
$$= -0.1250$$
$$\hat{\alpha}_D = [-0.17 + (-0.83) + (-0.34)]/4$$
$$= -1.34/4$$
$$= -0.3350$$

The main effects of the four treatments should total 0.

$$\hat{\alpha}_A + \hat{\alpha}_B + \hat{\alpha}_C + \hat{\alpha}_D = -0.6225 + 1.0825 - 0.1250 - 0.3350$$
$$= 1.0825 - 1.0825$$
$$= 0$$

The order effect ($\hat{\delta}$) is calculated by totaling the mean score for each combination of pairs and dividing by the number of ordered pairs (12).

$$\text{The order effect } (\hat{\delta}) = (-1.33 + 2.33 - 0.33 + 1.33 + 1.00 + 0.67 + 1.33 - 2.00 + 1.33 - 0.33 - 0.33 - 1.00)/12$$
$$= 2.67/12$$
$$= 0.22$$

NOTE: IF THE READER IS NOT FAMILIAR WITH THE ANALYSIS OF VARIANCE PROCEDURE, HE SHOULD REFER TO THE COMPLETE DESCRIPTION GIVEN UNDER THE MULTIPLE COMPARISONS TEST, PAGE 32.

Analysis of variance table

Source of variation	*df*	*ss*	*ms*	*F*
Main effects	3	40.49	13.50	9.85
Order effect	1	1.74	1.74	1.27
Error	32	43.77	1.37	
Total	36	86.00		

df — degrees of freedom
ss — sum of squares
ms — mean of squares
F — variance ratio

Sum of squares for main effects = number of judges × number of treatments × sum of the square of the main effects for each treatment.

$$= 6 \times 4 \times [(-0.6225)^2 + 1.0825^2 + (-0.1250)^2 + (-0.3350)^2]$$
$$= 24 \times 1.6871$$
$$= 40.49$$

Sum of squares, order effect = number of panelists × number of pairs × the square of the order effect ($\hat{\delta}$).

$$= 6 \times 6 \times 0.22^2$$
$$= 1.74$$

Sum of squares, total = The sum of the squares of each value (using the frequency of each score).

$$= 3^2(2) + 2^2(13) + 1^2(16) + 0^2(5)$$
$$= 86.00$$

According to Statistical Chart 3 in the Appendix, page 65, the *F* ratio for main effects is significant but the *F* ratio for order effect is not significant. Tukey's Test is used to determine which treatments are significantly different.

The standard error of the mean is calculated:

$$
\begin{aligned}
SE &= \sqrt{ms \text{ for error/number of judgments made on each sample}} \\
&= \sqrt{1.37/18} \\
&= \sqrt{.076111} \\
&= .2759
\end{aligned}
$$

The value from Chart 4, page 67, for four samples and 32 degrees of freedom is 3.83.

$$
\begin{aligned}
\text{Least significant difference} &= 3.83 \times 0.2759 \\
&= 1.0567
\end{aligned}
$$

Average preferences:

$\hat{\alpha}_A$	$\hat{\alpha}_B$	$\hat{\alpha}_C$	$\hat{\alpha}_D$
–0.6225	1.0825	–0.1250	–0.3350

In order of magnitude:

$\hat{\alpha}_B$	$\hat{\alpha}_C$	$\hat{\alpha}_D$	$\hat{\alpha}_A$
1.0825	–0.1250	–0.3350	–0.6225

To determine which samples are significantly more juicy, subtract the main effects of treatment from each other. Compare the difference with the least significant difference.

$$
\begin{aligned}
\hat{\alpha}_B - \hat{\alpha}_A &= 1.0825 + 0.6225 = 1.7050 > 1.0567 \\
\hat{\alpha}_B - \hat{\alpha}_D &= 1.0825 + 0.3350 = 1.4175 > 1.0567 \\
\hat{\alpha}_B - \hat{\alpha}_C &= 1.0825 + 0.1250 = 1.2075 > 1.0567 \\
\hat{\alpha}_C - \hat{\alpha}_A &= 0.1250 + 0.6225 = 0.4975 < 1.0567
\end{aligned}
$$

$\hat{\alpha}_B$	$\hat{\alpha}_C$	$\hat{\alpha}_D$	$\hat{\alpha}_A$
1.0825*a*	–0.1250*b*	–0.3350*b*	–0.6225*b*

Any two treatments not followed by the same letter are significantly different at the 5% level of significance. The conclusion is that Treatment B is significantly more juicy than the other three treatments.

This conclusion does not necessarily mean that the samples are dry. Paired comparisons do not give an absolute measure of quality but only a measure in relation to the samples with which it is compared. To obtain some idea of how juicy or how dry the samples are, the second question is asked. Scores from one to six are assigned to the ratings.

The mean score for each treatment is calculated to obtain an indication of dryness or juiciness.

Duo-trio test

In the duo-trio test, three samples are presented to the taster; one is labeled R (reference) and the other two are coded. One coded sample is identical with R and the other is different. The panelist is asked to identify the odd sample.

The duo-trio test has the same applications as the triangle test but is less efficient because the probability of selecting the correct sample by chance is 50%. This test is often used instead of the triangle test when tasting samples that have a strong flavor because less tasting is required.

It can be used in place of the paired comparisons test if no characteristic is specified. In the simple paired comparisons test the panelist is asked which sample has more of some specified characteristic, whereas in the duo-trio and the triangle tests the panelist bases his judgment on any difference he can detect.

QUESTIONNAIRE FOR
DUO-TRIO TEST

NAME________________________________DATE_____________________________

PRODUCT____

On your tray you have a marked control sample (R) and two coded samples. One sample is identical with R and the other is different. Which of the coded samples is different from R?

Samples	Check odd sample
432	______________________
701	______________________

Comments:

EXAMPLE. A duo-trio test was used to determine if methional could be detected when added to cheddar cheese in amounts of 0.125 ppm and 0.250 ppm. Each tray had a control sample marked R and two coded samples, one with methional added and one with no methional. The duo-trio test was used in preference to the triangle test because less tasting is required to form a judgment. This fact is important when tasting a substance with a lingering aftertaste, such as methional.

The test was performed on two successive days using eight judges. Each day the judges were presented with two trays. One tray contained a sample with 0.125 ppm methional and two control samples and the other contained a sample with 0.250 ppm methional and two control samples. A total of 16 judgments were made at each level. The results are shown in the following table.

Duo-trio test on cheddar cheese containing 0.125 and 0.250 ppm methional

	Level of methional added, ppm			
Judges	1st day		2nd day	
	0.125	0.250	0.125	0.250
1	X	R	R	R
2	R	R	R	R
3	X	R	X	R
4	R	X	X	R
5	R	R	R	R
6	X	R	X	X
7	R	R	R	R
8	R	R	R	R
Total	5	7	5	7

X = Wrong
R = Right

0.125 ppm = 10 out of 16 correct judgments
0.250 ppm = 14 out of 16 correct judgments

Consult Statistical Chart 2 in the Appendix, page 64, for 16 judges in a two-sample test. This chart shows that 14 correct judgments are significant at the 1% level, but 10 correct judgments are not significant, even at the 5% level.

The conclusion is that methional added to cheddar cheese can be detected at the 0.250 ppm level but not at the 0.125 ppm level.

Multiple comparisons test

In multiple comparisons tests, a known reference or standard sample is labeled R and presented to the panelist with several coded samples. The panelist is asked to compare each coded sample with the reference sample on the basis of some named characteristics.

The multiple comparison method may be used to examine effects of replacing or changing an ingredient, of packaging material, of changing a process, or of storage. This test can be used very efficiently to evaluate four or five samples at a time. Small differences between the sample and the control can be detected. It also gives information about the direction and magnitude of the difference.

QUESTIONNAIRE FOR MULTIPLE COMPARISONS TEST

NAME____________________________DATE________________________

You are receiving samples of ___peaches___ to compare for ___firmness___. You have been given a reference sample, marked R, with which you are to compare each sample. Taste each sample; determine whether it is firmer than, comparable to, or softer than the reference. Then mark the amount of difference that exists.

Sample number	426	591	497	874				
Firmer than R								
Equal to R								
Softer to R								

AMOUNT OF DIFFERENCE:

None								
Slight								
Moderate								
Much								
Extreme								

Comments:

EXAMPLE. A multiple comparisons test was conducted to compare the firmness of four brands of canned peaches. One brand of known quality was selected as the control and was labeled R. Each of nine judges received a tray containing one R sample and four coded samples. The control sample was also included as a coded sample. The judges were asked to evaluate these samples according to the previous score sheet.

Numerical scores were assigned to the ratings by the experimenter, with "no difference" equaling five, "extremely firmer than R" equaling nine, and "extremely softer than R" equaling one. Analysis of variance of the scores was conducted as shown below.

	Brands				
Judges	A	B	C	D	Total
1	5	4	4	2	15
2	8	3	5	2	18
3	6	4	4	1	15
4	3	3	4	2	12
5	5	2	3	1	11
6	7	2	5	1	15
7	4	3	2	1	10
8	7	4	1	1	13
9	4	3	2	2	11
Total	49	28	30	13	120

Analysis of variance

To complete an analysis of variance, certain calculations must be performed to determine the correction factor (CF), the sum of squares (*ss*), the degrees of freedom (*df*), the mean square (*ms*), and the variance ratio (or *F* value). A detailed explanation of these calculations using the results from the previous table follows.

Correction factor — The correction factor (CF) is calculated by squaring the total and dividing by the total number of responses.

$$\begin{aligned} CF &= (120)^2/(4 \times 9) \\ &= 14400/36 \\ &= 400 \end{aligned}$$

Sum of squares, samples — The sum of squares (*ss*) for samples is calculated by adding the squares of the total for each sample, dividing by the number of judgments for each sample, and then subtracting the correction factor.

$$\begin{aligned} ss\text{, samples} &= (49^2 + 28^2 + 30^2 + 13^2)/9 - CF \\ &= 4254/9 - 400 \\ &= 472.67 - 400 \\ &= 72.67 \end{aligned}$$

Sum of squares, judges — The sum of squares (*ss*) for judges is calculated by adding the squares of the total for each judge, dividing by the number of judgments made by each judge, and subtracting the correction factor.

$$ss\text{, judges} = (15^2 + 18^2 + 15^2 + 12^2 + 11^2 + 15^2 + 10^2 + 13^2 + 11^2)/4 - CF$$
$$= 1654/4 - 400$$
$$= 413.5 - 400$$
$$= 13.5$$

Sum of squares, total — The sum of squares (*ss*) for the total is calculated by adding the square of each judgment and subtracting the correction factor.

$$ss\text{, total} = (5^2 + 8^2 + 6^2 + 3^2 + 5^2 + 7^2 + 4^2 + 7^2 + 4^2 + 4^2 + 3^2 + 4^2 + 3^2 + 2^2 + 2^2 + 3^2 + 4^2 + 3^2 + 4^2 + 5^2 + 4^2 + 4^2 + 3^2 + 5^2 + 2^2 + 1^2 + 2^2 + 2^2 + 2^2 + 1^2 + 2^2 + 1^2 + 1^2 + 1^2 + 1^2 + 2^2) - CF$$
$$= 518 - 400$$
$$= 118$$

Sum of squares, error — To determine the sum of squares (*ss*) for error, subtract the *ss* values obtained from the sources of variation (judges, samples) from the *ss* for the total.

$$ss\text{, error} = 118 - 72.67 - 13.50$$
$$= 31.83$$

Degrees of freedom, samples — The degrees of freedom (*df*) for samples are calculated by subtracting one from the number of samples.

$$df\text{, samples} = 4 - 1$$
$$= 3$$

Degrees of freedom, judges — The degrees of freedom (*df*) for judges are calculated by subtracting one from the number of judges.

$$df\text{, judges} = 9 - 1$$
$$= 8$$

Degrees of freedom, total — The degrees of freedom (*df*) for total are calculated by subtracting one from the number of judgments.

df, total = 36 – 1
= 35

Degrees of freedom, error — The degrees of freedom (*df*) for error are determined by subtracting the *df* for the other sources from the *df* for the total.

df, error = 35 – 8 – 3
= 24

Mean square — The mean square (*ms*) for any variable is determined by dividing the sum of squares (*ss*) for each variable by its respective degrees of freedom (*df*).

ms, samples = 72.67/3
= 24.22

ms, judges = 13.5/8
= 1.69

ms, error = 31.83/24
= 1.33

Variance ratio, samples — The variance ratio or *F* value for samples is determined by dividing the *ms* for samples by the *ms* for error.

F, samples = 24.22/1.33
= 18.21

Variance ratio, judges — The *F* value for judges is determined by dividing the *ms* for judges by the *ms* for error.

F, judges = 1.69/1.33
= 1.27

Analysis of variance table

Source of variation	*df*	*ss*	*ms*	*F*
Samples	3	72.67	24.22	18.21**
Judges	8	13.50	1.69	1.27
Error	24	31.83	1.33	
Total	35	118.0		

critical value

To determine if the difference between the samples is significant, the calculated *F* value (18.21) is checked on Chart 3 of the Appendix on pages 65 and 66. With three degrees of freedom in the numerator and 24 degrees of freedom in the denominator, the variance ratio (*F* value) must exceed 3.01 to be significant at the 5% level (*) and must exceed 4.72 to be significant at the 1%

level (**). The value 18.21 is therefore significant at the 1% level (**). If the variance ratio is not significant, the conclusion is that there is no difference in firmness among the samples.

Since there is a significant difference among the samples, the ones that are different can be determined using Tukey's Test (Snedecor 1956).

Sample scores:	A	B	C	D
	49	28	30	13

Sample means: score/number of judgments for each sample

	A	B	C	D
	49/9	28/9	30/9	13/9
	5.44	3.11	3.33	1.44

The sample means are arranged according to magnitude.

A	C	B	D
5.44	3.33	3.11	1.44

The standard error (SE) of the sample mean is calculated by taking the square root of the *ms* for error divided by the number of judgments for each sample.

$$\begin{aligned} SE &= \sqrt{(1.33/9)} \\ &= \sqrt{0.1478} \\ &= 0.38 \end{aligned}$$

The least significant difference is determined using Chart 4 of the Appendix on page 67. In this case there were four treatments and 24 *df* for error so the value obtained from the table is 3.90. The least significant difference is calculated by multiplying the value obtained from the the table by the standard error of the mean.

$$\begin{aligned} \text{Least significant difference} &= 3.90 \times 0.38 \\ &= 1.48 \end{aligned}$$

Any two sample means that differ by 1.48 or more are significantly different at the 5% level.

The differences between the sample means are calculated by comparing the sample that has the highest mean with each of the others beginning with the lowest and going up until no significant difference is found.

A	C	B	D
5.44	3.33	3.11	1.44

$$A - D = 5.44 - 1.44 = 4.00 > 1.48$$
$$A - B = 5.44 - 3.11 = 2.33 > 1.48$$
$$A - C = 5.44 - 3.33 = 2.11 > 1.48$$

A is significantly firmer than the other three samples.

Next, test to see which samples the second sample, C, is different from:

$C - D = 3.33 - 1.44 = 1.89 > 1.48$
$C - B = 3.33 - 3.11 = 0.22 < 1.48$

C is significantly firmer than D. Finally, test to see if B differs from D:

$B - D = 3.11 - 1.44 = 1.67 > 1.48$

B is significantly firmer than D.

These results can be shown by using letters to indicate differences:

A	C	B	D
5.44*a*	3.33*b*	3.11*b*	1.44*c*

Any two means not followed by the same letter are significantly different at the 5% level. The conclusion is that brand A peaches are firmer than peaches from brands B, C, and D. Brands B and C are firmer than brand D.

Ranking

The panelist receives three or more coded samples and he is asked to rank them for the intensity of some specific characteristic. The results of a ranking test can be checked for significant differences by using tables prepared by Kramer et al. (1974).

Before these tables became available, the results were transformed from ranks to scores using Table XX from Fisher and Yates (1942).

The ranking method is rapid and allows the testing of several samples at once. It is generally used to screen one or two of the best samples from a group of samples rather than to test all samples thoroughly. No indication of the size of the differences between samples is obtained. Because samples are evaluated only in relation to each other, results from one set of ranks cannot be compared with the results from another set of ranks unless both contain only the same samples.

An example of the ranking method and a sample questionnaire follow.

QUESTIONNAIRE FOR
RANKING

NAME ______________________________ DATE ____________________

Rank these samples for sweetness. The sweetest sample is ranked first, the second sweetest sample is ranked second, the third sweetest sample is ranked third, and the least sweet sample is ranked fourth. Place the code numbers on the appropriate lines. Test the samples in the following order: 212 336 471 649

____________	____________	____________	____________
1	2	3	4

Comments:

EXAMPLE. A ranking test was used to compare the sweetness of a fruit drink made using four different sweetening agents. Eight judges ranked the four drinks using the score sheet above. The ranks given the samples by the judges are shown below.

	Samples			
Judges	A	B	C	D
1	4	2	1	3
2	4	3	1	2
3	3	1	2	4
4	3	2	1	4
5	4	1	2	3
6	4	3	1	2
7	4	2	1	3
8	4	1	2	3
Total	30	15	11	24

The results can be analyzed using two different methods.

Method I

The rank totals are compared with the values in Chart 5 of the Appendix on page 68. When there are four treatments and eight judges (reps), the tabular entries are 13–27: the lowest insignificant rank sum is 13 and the highest insignificant rank sum is 27. If one or more rank sums are lower than the upper left value in the block (13) or higher than the upper right value in the block (27), statistical significance at the 5% level of significance is indicated. Since one rank sum (11) is lower than 13 and one (30) is higher than 27, there is a significant difference among the treatments. All of the rank sums are now compared to the lower pair of values in the block to determine which samples are significantly low (sweeter) or high (less sweet). Sample C, which received a rank sum of 11, is significantly sweeter than Sample A, which received a rank sum of 30.

The conclusion is that Sample C is significantly sweeter than Sample A.

Method II

The ranks are converted to scores according to the method of Fisher and Yates (1942). Table 6, page 70, is used to determine the score given to each rank. The sample ranked first of four samples is given a value of +1.03; the second sample is +0.30; the third is –0.30; and the fourth is –1.03. When there is an odd number of samples the middle ranked sample receives a score of 0. If there were five ranked samples the scores would be:

first = 1.16
second = 0.50
third = 0
fourth = –0.50
fifth = –1.16

The ranks shown on page 38 are converted to scores to give the following table:

	Samples				
Judges	A	B	C	D	Total
1	−1.03	+0.30	+1.03	−0.30	0
2	−1.03	−0.30	+1.03	+0.30	0
3	−0.30	+1.03	+0.30	−1.03	0
4	−0.30	+0.30	+1.03	−1.03	0
5	−1.03	+1.03	+0.30	−0.30	0
6	−1.03	−0.30	+1.03	+0.30	0
7	−1.03	+0.30	+1.03	−0.30	0
8	−1.03	+1.03	+0.30	−0.30	0
Total	−6.78	+3.39	+6.05	−2.66	0
Mean	−0.85	+0.42	+0.76	−0.33	

The scores are then subjected to analysis of variance.

NOTE: IF THE READER IS NOT FAMILIAR WITH THE ANALYSIS OF VARIANCE PROCEDURE, HE SHOULD REFER TO THE COMPLETE DESCRIPTION GIVEN UNDER THE MULTIPLE COMPARISONS TEST, PAGE 32.

Analysis of variance

Correction factor = Total^2/number of responses

CF = $0^2/32$
= 0

Sum of squares, samples = (Sum of the squares of the total for each sample/number of judgments for each sample) – CF
= $[(-6.78)^2 + (3.39)^2 + (16.05)^2 + (-2.66)^2]/8 - CF$
= $(45.97 + 11.49 + 36.60 + 7.07)/8 - CF$
= $101.14/8 - 0$
= 12.64

Sum of squares, judges = (Sum of the square of the total for each judge/number of judgments by each judge) – CF
= $0^2/4 - 0$
= 0

Sum of squares, total = sum of the square of each judgment – CF
= $(-1.03)^2 + (-1.03)^2 + \ldots + (-0.30)^2 - CF$
= $18.41 - 0$
= 18.41

Analysis of variance table

Source of variation	*df*	*ss*	*ms*	*F*
Samples	3	12.64	4.21	15.59**
Judges	7	0	0	0
Error	21	5.77	0.27	
Total	31	18.41		

With three degrees of freedom in the numerator and 21 degrees of freedom in the denominator, the variance ratio (*F* value) must exceed 3.07 to be significant at the 5% level(*) (Chart 3, page 65) and it must exceed 4.87 to be significant at the 1% level(**). The calculated *F* value is 15.59. There is a significant difference at the 1% level(**).

To determine which samples are significantly different from one another, Tukey's Test is used.

	A	B	C	D
Sample scores:	–6.78	+3.39	+6.05	–2.66
Sample means:	–0.85	+0.42	+0.76	–0.33

The sample means are arranged according to magnitude.

C	B	D	A
+0.76	+0.42	–0.33	–0.85

The standard error of the sample mean is calculated.

$$SE = \sqrt{(\text{ms, error/number of judgments for each sample})}$$
$$= \sqrt{0.27/8}$$
$$= \sqrt{0.03375}$$
$$= 0.18$$

The least significant difference is determined using Chart 4 of the Appendix, page 67. In this case there were four treatments and 21 *df* for error so the value obtained from the table is 3.96. The least significant difference is calculated by multiplying 3.96 by the standard error of the mean.

$$\text{Least significant difference} = 3.96 \times 0.18$$
$$= 0.71$$

Any two samples that differ by 0.71 or more are significantly different at the 5% level.

The differences between the sample means are calculated.

$C - A = 0.76 + 0.85 = 1.61 > 0.71$
$C - D = 0.76 + 0.33 = 1.09 > 0.71$
$C - B = 0.76 - 0.42 = 0.34 < 0.71$
$B - A = 0.42 + 0.85 = 1.27 > 0.71$
$B - D = 0.42 + 0.33 = 0.75 > 0.71$
$D - A = 0.33 + 0.85 = 0.52 < 0.71$

The results can be shown by using letters to indicate differences:

C	B	D	A
+0.76*a*	+0.42*a*	–0.33*b*	–0.85*b*

Any two means not followed by the same letter are significantly different at the 5% level. The conclusion is that samples C and B are significantly sweeter than samples D and A.

Methods I and II give slightly different, but consistent results. The analysis of variance with Tukey's Test (Method II) is a more detailed method yielding more precise results. Since ranking is generally used to screen out the best or the worst from several samples, the results obtained by Method I are probably sufficient.

Scoring

Coded samples are evaluated for the intensity of some specified characteristic (Fig. 5). The panelist records his judgment on a graduated scale. The intervals on the scale can be labeled with numbers or with descriptive terms.

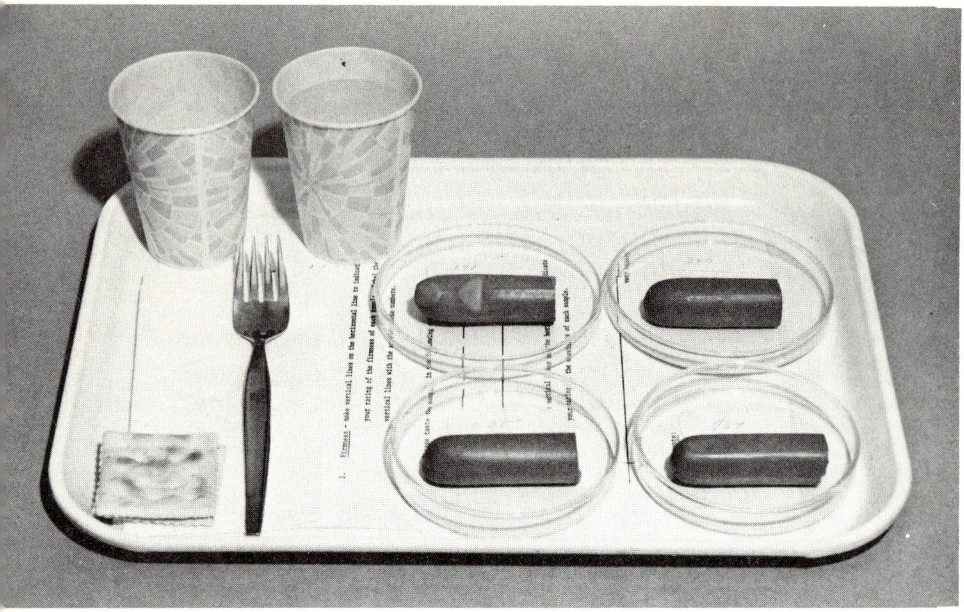

Fig. 5. Tray prepared for scoring test.

From the scores, the size and the direction of the differences between samples are evident. From the statistical analysis the experimenter determines whether or not the differences are significant.

The descriptive terms on the scale must be carefully selected and the panelists trained so that they agree on the meaning of the terms. Objective terms, such as "very hard," rather than preference terms, such as "much too hard," must be used. The panelists are not typical consumers and their likes and dislikes are not considered.

To use scoring effectively, all the panelists must be evaluating the same characteristic. This is no problem when a simple characteristic like sweetness is involved but when a complex characteristic is being evaluated, like the texture of French fries for example, the panelists might not have the same concept.

The inclusion of standards at various points in the scale will help minimize panel variability. These standards act as anchors in counteracting the tendency of scales to drift in meaning with time. This instability is a marked disadvantage when scoring is used in storage stability studies over an extended period.

An example of the scoring method and a sample questionnaire follow.

QUESTIONNAIRE FOR
SCORING

NAME ______________________ DATE ______________________

Evaluate these samples for bitterness. Indicate the amount of bitterness in each sample on the scales below.

419	172	386
_____ not bitter	_____ not bitter	_____ not bitter
_____ trace of bitterness	_____ trace of bitterness	_____ trace of bitterness
_____ slightly bitter	_____ slightly bitter	_____ slightly bitter
_____ bitter	_____ bitter	_____ bitter
_____ very bitter	_____ very bitter	_____ very bitter
_____ extremely bitter	_____ extremely bitter	_____ extremely bitter

Comments:

EXAMPLE. The scoring method was used to determine if there was a difference in bitterness in cheddar cheese made using three different milk coagulating enzymes. Samples of cheese from each treatment were coded and presented to 12 judges for evaluation. The order in which the three samples were tasted was balanced so that each possible order was used twice: ABC, ACB, BAC, BCA, CAB, CBA.

The ratings assigned by the judges were given numerical values, ranging from 0 points for "not bitter" to 5 points for "extremely bitter." The results are shown in the following table.

Judges	Samples			Total
	A	B	C	
1	3	0	1	4
2	2	2	2	6
3	3	1	2	6
4	1	1	0	2
5	3	1	3	7
6	2	1	1	4
7	3	2	2	7
8	2	0	1	3
9	3	1	2	6
10	4	2	3	9
11	1	1	0	2
12	2	2	2	6
Total	29	14	19	62
Mean	2.42	1.17	1.58	

The results were submitted to analysis of variance.

NOTE: IF THE READER IS NOT FAMILIAR WITH THE ANALYSIS OF VARIANCE, HE SHOULD REFER TO THE DETAILED DESCRIPTION GIVEN UNDER THE MULTIPLE COMPARISONS METHOD, PAGE 32.

Analysis of variance

Correction factor (CF) $= 62^2/36$
$= 106.78$

Sum of squares, samples $= (29^2 + 14^2 + 19^2)/12 - \text{CF}$
$= 1398/12 - \text{CF}$
$= 116.50 - 106.78$
$= 9.72$

Sum of squares, judges $= (4^2 + 6^2 + 6^2 + \ldots + 6^2)/3 - \text{CF}$
$= 372/3 - \text{CF}$
$= 124 - 106.78$
$= 17.22$

Sum of squares, total $= (3^2 + 2^2 + 3^2 + 1^2 + \ldots + 2^2) - \text{CF}$
$= 142 - 106.78$
$= 35.22$

Analysis of variance table

Source of variation	*df*	*ss*	*ms*	*F*
Samples	2	9.72	4.86	12.79**
Judges	11	17.22	1.57	4.13**
Error	22	8.28	0.38	
Total	35	35.22		

The *F* value for samples is 12.79. According to Chart 3 in the Appendix, page 65, if the *F* value exceeds 3.44 there is a significant difference at the 5% level (*) and if it exceeds 5.72 there is a significant difference at the 1% level (**). There is a significant difference at the 1% level so the value is marked with two asterisks.

Tukey's Test is used to determine which samples are significantly different.

The standard error of the mean is calculated.

$$\begin{aligned} SE &= \sqrt{0.38/12} \\ &= \sqrt{.0317} \\ &= 0.18 \end{aligned}$$

The value from Chart 4 of the Appendix, page 67, for three samples and 22 degrees of freedom is 3.58.

$$\begin{aligned} \text{Least significant difference} &= 3.58 \times 0.18 \\ &= 0.64 \end{aligned}$$

	A	B	C
Sample scores:	29	14	19
Sample means:	2.42	1.17	1.58

The sample means are arranged according to magnitude.

A	C	B
2.42	1.58	1.17

Each mean is compared with the others to see if the difference is 0.64 or more.

$A - B = 2.42 - 1.17 = 1.25 > 0.64$

$A - C = 2.42 - 1.58 = 0.84 > 0.64$

$C - B = 1.58 - 1.17 = 0.41 < 0.64$

The results are shown using letters to indicate differences:

A	C	B
2.42*a*	1.58*b*	1.17*b*

Any two values not followed by the same letter are significantly different at the 5% level. Sample A is significantly more bitter than samples C and B.

The analysis of variance table indicates that there is a significant difference among the judges. The tabular values from Chart 3 of the Appendix, page 65, indicate that the *F* value must exceed 2.23 to be significant at the 5% level and exceed 3.12 to be significant at the 1% level. Tukey's Test is used to see which judges differ significantly.

The standard error of the mean is calculated.

$$SE = \sqrt{0.38/3}$$
$$= \sqrt{0.1267}$$
$$= 0.36$$

The value from Chart 4 of the Appendix, page 67, for 12 judges and 22 degrees of freedom is 5.20.

Least significant difference = 5.20 × 0.36
= 1.87

Judges											
1	2	3	4	5	6	7	8	9	10	11	12
Totals											
4	6	6	2	7	4	7	3	6	9	2	6
Means											
1.33	2.0	2.0	0.67	2.33	1.33	2.33	1.0	2.0	3.0	0.67	2.0

The means are arranged in order of magnitude.

Judges											
10	5	7	2	3	9	12	1	6	8	4	11
Means											
3.0	2.33	2.33	2.0	2.0	2.0	2.0	1.33	1.33	1.0	0.67	0.67

Each mean is compared with the others to see if the difference is 1.87 or more.

$J10 - J11 = 3.0 - 0.67 = 2.33 > 1.87$
$J10 - J4 = 3.0 - 0.67 = 2.33 > 1.87$
$J10 - J8 = 3.0 - 1.0 = 2.0 > 1.87$
$J10 - J6 = 3.0 - 1.33 = 1.67 < 1.87$
$J5 - J11 = 2.33 - 0.67 = 1.66 < 1.87$

The results are shown using letters to indicate differences:

J10	J5	J7	J2	J3	J9
3.0*a*	2.33*ab*	2.33*ab*	2.0*ab*	2.0*ab*	2.0*ab*
J12	J1	J6	J8	J4	J11
2.0*ab*	1.33*ab*	1.33*ab*	1.0*b*	0.67*b*	0.67*b*

Judge 10 scored significantly higher than judges 11, 4, and 8.

When the scores of these judges are examined, it is evident that although judge 10 scored high, his scores are consistent with the means for the panel: A received the highest score and B received the lowest scores. Judge 8 gave low scores but again the scores are consistent with the panel mean. Judges 4 and 11, however, did not discriminate as well. They gave A and B the same score.

The fact that a judge assigns scores that are higher or lower than those of the rest of the panel is not as important as whether he is able to discriminate and recognize differences.

Ratio-scaling

A scaling method that is commonly used in physics is ratio-scaling. Scales of weights and distances are examples. A distance of 40 km is twice as long as a distance of 20 km. Ratios of measurements can be calculated. Ratio-scaling is also used in sensory evaluation. The ratio measurements are usually constructed by a procedure called magnitude estimation (Moskowitz et al. 1972). The panelist is given a series of samples that vary in one characteristic, such as hardness. He is instructed to assign a number, such as 50, to the first sample and rate each sample in relation to the first. If the second sample seems twice as hard as the first, he assigns to it the value 100; if it seems half as hard he rates the hardness at 25.

The number assigned to the first sample may be chosen beforehand by the experimenter or the panelist may select any number. No high or low limits are specified. One aim of the method of magnitude estimation is to have the panelist measure in the same way as nature measures, with ratio scale values that have no arbitrarily limited endpoints (Moskowitz 1974).

Because panelists will always vary in their use of numbers, the data are transformed so that the geometric mean for each panelist across all samples is 1.00. The logarithms of transformed data is submitted to analysis of variance.

DESCRIPTIVE SENSORY ANALYSIS

A group of highly trained panelists examine the flavor or texture of a product to provide a detailed descriptive evaluation of it. The most commonly known descriptive methods are the flavor profile (Caul 1957) and the texture profile (Brandt et al. 1963).

The flavor profile is the description of the flavor and aroma of a food product. The description names the perceptible factors, the intensity of each factor, the order in which the factors are perceived, aftertaste, and overall impression.

The texture profile is the description of the textural characteristics perceived in a food product, the intensity of each factor, and the order in which they are perceived. Mechanical characteristics are described qualitatively and quantitatively; geometrical characteristics are described qualitatively and semiquantitatively; the description of fat and moisture characteristics depends on the product being studied. Hardness, fracturability, chewiness, gumminess, adhesiveness, and viscosity are mechanical properties. Examples of geometrical properties are grittiness, coarseness, and fibrousness.

Descriptive analysis is a valuable tool in difference testing and in product development work. It provides a complete description of sample differences and guides the product developer in modifying product characteristics to meet consumer demands.

The training of profile panels requires considerable time and the members must possess a high degree of motivation and interest. Once trained, however, the panel can provide thorough and reliable descriptions of products in a short time. Since the descriptive panel members work together as a group, forceful members could have undue influence on the other panelists and hence, change the results.

More information on these methods can be found in Caul (1957) and Brandt et al. (1963).

Descriptive analysis with scaling

A method of sensory evaluation called quantitative descriptive analysis was developed at the Stanford Research Institute (Stone et al. 1974). During preliminary sessions the sensory properties of the product are identified by a trained panel. Samples are made up to illustrate the different properties so that the panel agrees on the meaning of each term used. During these sessions the judges work together as a group and discussion is encouraged. The panel members decide the sensory properties that should be evaluated and they also develop the language to be used. During evaluation sessions the panelists work individually.

The scale used is an interval scale consisting of a horizontal line *6 in.* (15 cm) long with anchor points *0.5 in.* (1.3 cm) from each end and usually, but not necessarily, having a mid point. Each anchor point is usually labeled with a word or expression. A separate line is used for each sensory property to be evaluated.

Each judge records his evaluation by making a vertical line across the horizontal line at the point that best reflects his perception of the magnitude of that property.

After the panelists have completed their judgments, the experimenter superimposes a grid dividing the line into 60 units, to assign a number between 0 and 60 to each rating. These values are then tabulated.

For details regarding analysis of the data and interpretation of results, see Stone et al. (1974).

Structured scales such as those used in scoring are based on assumptions that may or may not hold true. It is assumed that each word phrase on the scale has the same meaning to each judge and that the terms on the scale represent equal sensory intervals. For instance, it is assumed that the distance between "slightly bitter" and "bitter" in the example on page 43 is the same as the distance between "very bitter" and "extremely bitter". This may or may not be the case.

Unstructured scales, with verbal anchors at the ends only, eliminate the problem of unequal intervals that is associated with structured scales. However, carry over of ratings from one experiment to another is difficult (Hall 1958).

Descriptive analysis with unstructured scales is used at the Food Research Institute. The panel works together as a group to identify important characteristics of the product to be evaluated. The terms to be used at the anchor points of the lines are agreed upon. Data are collected in the way described by Stone et al. (1974). After the data are tabulated, they are analyzed by analysis of variance in the same way as the data from the scoring of multiple comparisons methods.

A sample questionnaire and an example of descriptive analysis with scaling method follow.

QUESTIONNAIRE FOR DESCRIPTIVE ANALYSIS WITH SCALING

NAME ______________________________ DATE ____________________

Please evaluate the firmness and chewiness of these sample of wieners.

1. Firmness — make vertical lines on the horizontal line to indicate your rating of the firmness of each sample. Label each vertical line with the code number of the sample it represents.

Please taste the samples in the following order:

572   681   437   249

very soft   very firm

2. Chewiness — make vertical lines on the horizontal line to indicate your rating of the chewiness of each sample. Label each vertical line with the code number of the sample it represents.

very mushy   very rubbery

Comments:

EXAMPLE. This method was used when studying the texture of wieners. During preliminary testing the judges determined that firmness and chewiness were the most important textural characteristics of wieners. Four brands of wieners were evaluated by eight judges using the preceding score sheet. The anchor words were agreed upon by the judges during preliminary sessions.

Numerical values were given to the ratings by measuring the distance of the judges' marks from the left end of the line in units of *0.1 in.* (0.25 cm).

Analysis of variance was conducted on the numerical values for each characteristic.

Firmness

Judges	Brands A	B	C	D	Total
1	1.9	2.1	3.4	1.1	8.5
2	4.1	2.4	5.1	3.0	14.6
3	4.6	3.2	5.3	1.8	14.9
4	2.3	5.3	5.3	1.3	14.2
5	1.5	4.7	5.3	0.6	12.1
6	2.0	3.3	4.7	1.8	11.8
7	2.0	3.5	5.2	2.7	13.4
8	2.1	4.8	5.4	1.7	14.0
Total	20.5	29.3	39.7	14.0	103.5
Mean	2.6	3.7	5.0	1.8	

NOTE: IF THE READER IS NOT FAMILIAR WITH THE ANALYSIS OF VARIANCE PROCEDURE, HE SHOULD REFER TO THE DETAILED DESCRIPTION GIVEN UNDER THE MULTIPLE COMPARISONS METHOD, PAGE 32.

Analysis of variance

Correction factor (CF) = Total ^{2}/number of responses
= $103.5^2/32$
= 10712.25/32
= 334.76

Sum of squares, samples = (Sum of the squares of the total for each sample/number of judgments for each sample) – CF
= $(20.5^2 + 29.3^2 + 39.7^2 + 14.0^2)/8$ – CF
= 3050.83/8 – CF
= 381.35 – 334.76
= 46.59

Sum of squares, judges = (Sum of the squares of the total for each judge/number of judgments by each judge) – CF
= $(8.5^2 + 14.6^2 + 14.9^2 + \ldots + 14.0^2)/4$ – CF
= 1370.27/4 – CF
= 342.57 – 334.76
= 7.81

Sum of squares, total = Sum of the squares of each judgment − CF

$= (1.9^2 + 4.1^2 + 4.6^2 + \ldots + 1.7^2) - CF$

$= 407.35 - 334.76$

$= 72.59$

Analysis of variance table

Source of variation	*df*	*ss*	*ms*	*F*
Samples	3	46.59	15.53	17.85**
Judges	7	7.81	1.01	1.16
Error	21	18.19	0.87	
Total	31	72.59		

The *F* value for samples is 17.85. According to Chart 3 in the Appendix, page 65, if the *F* value exceeds 3.07 there is a significant difference at the 5% level (*) and if it exceeds 4.87 there is a significant difference at the 1% level (**). There is a significant difference at the 1% level for samples so the value is marked with two asterisks. There is no significant difference for judges.

Tukey's Test is used to determine which samples are significantly different.

The standard error of the mean is calculated.

$SE = \sqrt{0.87/8}$

$= \sqrt{.1088}$

$= 0.33$

The value from Chart 4 in the Appendix, page 67, for four samples and 21 degrees of freedom is 3.95.

Least significant difference $= 3.95 \times 0.33$

$= 1.30$

	A	B	C	D
Sample scores:	20.5	29.3	39.7	[illegible]
Sample means:	2.6	3.7	5.0	[illegible]

The sample means are arranged according to magnitude.

C	B	A	D
5.0	3.7	2.6	1.8

Each mean is compared with the others to see if the difference is 1.3 or more.

$C - D = 5.0 - 1.8 = 3.2 > 1.3$

$C - A = 5.0 - 2.6 = 2.4 > 1.3$

$C - B = 5.0 - 3.7 = 1.3 = 1.3$

Sample C is significantly different from all the others.
B – D = 3.7 – 1.8 = 1.9 > 1.3
B – A = 3.7 – 2.6 = 1.1 < 1.3

Sample B is significantly different from sample D but not from sample A.
A – D = 2.6 – 1.8 = 0.8 < 1.3

Samples A and D are not significantly different.

The results are shown using letters to indicate differences.

C	B	A	D
5.0*a*	3.7*b*	2.6*bc*	1.8*c*

Any two values not followed by the same letter are significantly different at the 5% level. Sample C is significantly firmer than all the other, and sample B is significantly firmer than sample D.

Chewiness

	Brands				
Judges	A	B	C	D	Total
1	2.2	1.9	3.2	0.2	7.5
2	2.5	1.0	4.2	2.0	9.7
3	3.3	3.0	4.0	1.5	11.8
4	2.6	5.4	4.2	1.3	13.5
5	2.6	4.3	5.1	0.1	12.1
6	2.2	3.4	4.7	1.9	12.2
7	2.4	3.6	5.5	3.2	14.7
8	1.6	3.9	5.7	1.8	13.0
Total	19.4	26.5	36.6	12.0	94.5
Mean	2.4	3.3	4.6	1.5	

Analysis of variance

Correction factor (CF) $= 94.5^2/32$
$= 8930.25/32$
$= 279.07$

Sum of squares, samples $= (19.4^2 + 26.5^2 + 36.6^2 + 12.0^2)/8 - CF$
$= 2562.17/8 - CF$
$= 320.27 - 279.07$
$= 41.20$

Sum of squares, judges $= (7.5^2 + 9.7^2 + 11.8^2 + \ldots + 13.0^2)/4 - CF$
$= 1152.17/4 - CF$
$= 288.04 - 279.07$
$= 8.97$

Sum of squares, total $= (2.2^2 + 2.5^2 + 3.3^2 + \ldots + 1.8^2) - CF$
$= 347.09 - 279.07$
$= 68.02$

Analysis of variance table

Source of variation	*df*	*ss*	*ms*	*F*
Samples	3	41.20	13.73	16.15**
Judges	7	8.97	1.28	1.51
Error	21	17.85	0.85	
Total	31	68.02		

The *F* value for samples is 16.15. According to Chart 3 of the Appendix, page 65, if the *F* value exceeds 3.07 there is a significant difference at the 5% level (*) and if it exceeds 4.87 there is a significant difference at the 1% level (**). There is a significant difference at the 1% level for samples so the value is marked with two asterisks. There is no significant difference for judges.

Tukey's Test is used to determine which samples are significantly different. The standard error of the mean is calculated.

$$SE = \sqrt{0.85/8}$$
$$= \sqrt{0.1062}$$
$$= 0.32$$

The value from Chart 4 of the Appendix, page 67, for four samples and 21 degrees of freedom is 3.95.

Least significant difference $= 3.95 \times 0.32$
$= 1.3$

	A	B	C	D
Sample scores:	19.4	26.5	36.6	12.0
Sample means:	2.4	3.3	4.6	1.5

The sample means are arranged according to magnitude.

C	B	A	D
4.6	3.3	2.4	1.5

Each mean is compared with the others to see if the difference is 1.3 or more.

$C - D = 4.6 - 1.5 = 3.1 > 1.3$
$C - A = 4.6 - 2.4 = 2.2 > 1.3$
$C - B = 4.6 - 3.3 = 1.3 = 1.3$ (least significant difference)
$B - D = 3.3 - 1.5 = 1.8 > 1.3$
$B - A = 3.3 - 2.4 = 0.9 < 1.3$
$A - D = 2.4 - 1.5 = 0.9 < 1.3$

The results are shown using letters to indicate differences:

C	B	A	D
4.6*a*	3.3*b*	2.4*bc*	1.5*c*

Any two values not followed by the same letter are significantly different at the 5% level. Sample C is significantly more chewy than all the others. Sample B is more chewy than sample D.

PREFERENCE TESTS

Preference tests include the paired comparison test, the hedonic scale, and ranking. Sample questionnaires and examples of these methods follow.

Paired comparison test

The paired comparison test used in preference testing is similar to that used for difference testing. When testing preferences, the panelist is presented with two coded samples (Fig. 6) and is asked which he prefers, such as in the next questionnaire.

Fig. 6. Tray prepared for paired comparison test.

QUESTIONNAIRE FOR
PAIRED COMPARISON TEST

NAME ______________________ DATE ______________________
PRODUCT ____

Taste the two coded samples in the following order: 317 225

Which of these two samples do you prefer? ________________

Comments:

EXAMPLE. To determine which of two spice combinations was preferred in a drop cookie, a paired comparison preference test was used. Fifty panelists compared the two cookies. Twenty-five panelists tasted a cookie from treatment A first and the other 25 tasted a cookie from treatment B first.

Thirty-five of the fifty panelists preferred the cookies from treatment B. According to statistical Chart 2 in the Appendix, page 64, this indicates a significant preference at the 1% level. If 33 persons had preferred one cookie over the other, it would have been significantly preferred at the 5% level.

Hedonic Scale

The most commonly used scale for preference testing is the nine-point hedonic scale. The term "hedonic" is defined as "having to do with pleasure." It should only be used in connection with scales in which the panelist expresses his degree of liking or disliking, such as in the following sample questionnaire.

QUESTIONNAIRE FOR HEDONIC SCALE

NAME ______________________________ DATE ______________________

PRODUCT ____

Taste these samples and check how much you like or dislike each one.

148	914	579
_____ like extremely	_____ like extremely	_____ like extremely
_____ like very much	_____ like very much	_____ like very much
_____ like moderately	_____ like moderately	_____ like moderately
_____ like slightly	_____ like slightly	_____ like slightly
_____ neither like nor dislike	_____ neither like nor dislike	_____ neither like nor dislike
_____ dislike slightly	_____ dislike slightly	_____ dislike slightly
_____ dislike moderately	_____ dislike moderately	_____ dislike moderately
_____ dislike very much	_____ dislike very much	_____ dislike very much
_____ dislike extremely	_____ dislike extremely	_____ dislike extremely

Comments:

EXAMPLE. The ratings for each sample are given numerical values ranging from like extremely (9) to dislike extremely (1). The results are analyzed by analysis of variance, which is described on page 32.

If only two treatments are evaluated, the mean scores received by each can be compared using the *t*-test.

The scores are tabulated:

Judges	Samples A	Samples B	Difference
1	8	6	2
2	7	7	0
3	7	6	1
4	8	7	1
5	6	6	0
6	7	6	1
7	7	7	0
8	8	7	1
9	6	7	−1
10	7	7	0
Total	71	66	5
Mean	7.1	6.6	0.5

T-test

Calculate $\bar{d}$ (average difference):

$$\bar{d} = \text{mean for A} - \text{mean for B} = 7.1 - 6.6 = 0.5$$

Calculate S:

$$S = \sqrt{\frac{\Sigma d_i^2 - [(\Sigma d)^2/n]}{n-1}}$$

Σd^2 = sum of the square of each difference
$= 2^2 + 0^2 + 1^2 + \ldots + 0^2$
$= 9$

$(\Sigma d)^2$ = the sum of the differences, squared
$= 5^2$
$= 25$

n = number of pairs
$= 10$

$$S = \sqrt{\frac{9 - 25/10}{9}}$$

$$= \sqrt{\frac{9 - 2.5}{9}}$$

$$= \sqrt{\frac{6.5}{9}}$$

$$= \sqrt{0.7222}$$

$$= 0.85$$

Find the t value from Statistical Chart 7 in the Appendix on page 71 under the column headed 0.050 (for 5% level of significance). The df is the number of pairs minus one.

$$df = 10 - 1$$
$$= 9$$
$$t \text{ value} = 2.262$$

The samples are significantly different if $\frac{\bar{d}}{S/\sqrt{n}} > t$

$$\frac{\bar{d}}{S/\sqrt{n}} = \frac{0.5}{0.85/\sqrt{10}}$$
$$= \frac{0.5}{0.85/3.16}$$
$$= \frac{0.5}{0.27}$$
$$= 1.85$$

1.85 is not greater than the t value, 2.262.

The conclusion is that there is no significant difference in the cookies from treatment A and the cookies from treatment B.

Ranking

When ranking for preference, the panelist is presented with coded samples to rank in order of preference.

QUESTIONNAIRE FOR
RANKING

NAME ______________________ DATE ____________________

PRODUCT ______________________________________

Please rank these samples for preference. Rank the sample you like best as first and the sample you like least as fourth.

Taste the samples in the following order: 817 462 149 535

First ____________

Second ____________

Third ____________

Fourth ____________

Comments:

The results can be analyzed according to Method I or Method II, which are described under Difference Test—Ranking, page 38.

REFERENCES

American Society for Testing and Materials. 1968. Manual on sensory testing methods. ASTM STP 434. American Society for Testing and Materials, Philadelphia, Pa. 77 pp.

Amerine, M. A., Pangborn, R. M., and Roessler, E. B. 1965. Principles of sensory evaluation of food. Academic Press, New York, N.Y. 602 pp.

Bradley, R. A., and Terry, M. E. 1952. Rank analysis of incomplete block designs. I. The method of paired comparisons. Biometrika 39:324-345.

Brandt, M. A., Skinner, E. B., and Coleman, J. A. 1963. Texture profile method. J. Food Sci. 28:404-409.

Caul, J. F. 1957. The profile method of flavour analysis. Pages 1-40 *in* Advances in food research. Vol. 7.

Cochran, W. G., and Cox, G. M. 1957. Experimental designs. 2nd ed. Wiley, New York, N.Y. 611 pp.

Dietrick, C. E. 1968. A computer program for paired comparisons with ties. Can. Inst. Food Technol. J. 1(4):170.

Ellis, B. H. 1961. A guide book for sensory testing. Continental Can Co., Chicago, Ill. 55 pp.

Fisher, R. A., and Yates, F. 1949. Statistical tables for biological agricultural and medical research. 3rd ed., Oliver and Boyd Ltd., London, England. 112 pp.

Hall, R. L. 1958. Flavor study approaches at McCormick and Co., Inc. Pages 224-240 *in* Flavor research and food acceptance. Reinhold, New York, N.Y.

Kefford, J. F., and Christie, E. M. 1960. Sensory tests for colour, flavour, and texture. CSIRO Food Preserv. Q. 20:47-56.

Klemmer, E. T. 1968. Psychological principles of sensory evaluation. Page 51 *in* Basic principles of sensory evaluation, ASTM STP 433. American Society for Testing and Materials, Philadelphia, Pa.

Kramer, A., Kahan, G., Cooper, D., and Papavasilian, A. 1974. A non-parametric ranking method for the statistical evaluation of sensory data. Chemical Senses and Flavor 1:121-133.

Kroll, B. J., and Pilgrim, F. J. 1961. Sensory evaluation of accessory foods with and without carriers. J. Food Sci. 26:122-124.

Mackey, A. O., and Valassi, K. 1956. The discernment of primary tastes in the presence of different food textures. Food Technol. 10:238-240.

Marshall, S. G., and Vaisey, M. 1972. Sweetness perception in relation to some textural characteristics of hydrocolloid gels. J. Texture Stud. 3:173-185.

Moskowitz, H. R. 1974. Sensory evaluation by magnitude estimation. Food Technol. 28(11):16-21.

Moskowitz, H. R., Drake, B., and Akesson, C. 1972. Psychological measures of texture. J. Texture Stud. 3:135-145.

Pettit, L. A. 1958. Informational bias in flavor preference testing. Food Technol. 12(1):12-14.

Roessler, E. B., Warren, J., and Guymon, J. F. 1948. Significance in triangular taste tests. Food Res. 13:503-505.

Roessler, E. B., Baker, G. A., and Amerine, M. A. 1956. One-tailed and two-tailed tests in organoleptic comparisons. Food Res. 21:117-121.

Scheffé, H. 1952. An analysis of variance for paired comparisons. J. Am. Stat. Assoc. 47:381-400.

Snedecor, G. W. 1956. Statistical methods. 5th ed. Iowa State College Press, Iowa State College, Ames, Iowa. 534 pp.

Stone, H., and Oliver, S. 1966. Effect of viscosity on the detection of relative sweetness intensity of sucrose solutions. J. Food Sci. 31:129-134.

Stone, J., Sidel, J., Oliver, S., and Woolsey, A. 1974. Sensory evaluation by quantitative descriptive analysis. Food Technol. 28(11):24-34.

STATISTICAL CHART 1

Triangle test, difference analysis

Number of tasters	Number of correct answers necessary to establish level of significance			Number of tasters	Number of correct answers necessary to establish level of significance		
	* 5%	** 1%	*** 0.1%		* 5%	** 1%	*** 0.1%
7	5	6	7	57	27	29	31
8	6	7	8	58	27	29	32
9	6	7	8	59	27	30	32
10	7	8	9	60	28	30	33
11	7	8	9	61	28	30	33
12	8	9	10	62	28	31	33
13	8	9	10	63	29	31	34
14	9	10	11	64	29	32	34
15	9	10	12	65	30	32	35
16	10	11	12	66	30	32	35
17	10	11	13	67	30	33	36
18	10	12	13	68	31	33	36
19	11	12	14	69	31	34	36
20	11	13	14	70	32	34	37
21	12	13	15	71	32	34	37
22	12	14	15	72	32	35	38
23	13	14	16	73	33	35	38
24	13	14	16	74	33	36	39
25	13	15	17	75	34	36	39
26	14	15	17	76	34	36	39
27	14	16	18	77	34	37	40
28	15	16	18	78	35	37	40
29	15	17	19	79	35	38	41
30	16	17	19	80	35	38	41
31	16	18	19	81	36	38	41
32	16	18	20	82	36	39	42
33	17	19	20	83	37	39	42
34	17	19	21	84	37	40	43
35	18	19	21	85	37	40	43
36	18	20	22	86	38	40	44
37	18	20	22	87	38	41	44
38	19	21	23	88	39	41	44
39	19	21	23	89	39	42	45
40	20	22	24	90	39	42	45
41	20	22	24	91	40	42	46
42	21	22	25	92	40	43	46
43	21	23	25	93	40	43	46
44	21	23	25	94	41	44	47
45	22	24	26	95	41	44	47
46	22	24	26	96	42	44	48
47	23	25	27	97	42	45	48
48	23	25	27	98	42	45	49
49	23	25	28	99	43	46	49
50	24	26	28	100	43	46	49
51	24	26	29	200	80	84	89
52	25	27	29	300	117	122	127
53	25	27	29	400	152	158	165
54	25	27	30	500	188	194	202
55	26	28	30	1000	363	372	383
56	26	28	31	2000	709	722	737

STATISTICAL CHART 2

Two-sample test

Number of judgments	Two-tail tests: Minimum agreeing judgments necessary to establish significant differentiation			One-tail tests: Minimum correct answers necessary to establish significant differentiation		
	Probability level			Probability level		
	5%	1%	0.1%	5%	1%	0.1%
5	....	....	....	5	....	....
6	....	....	....	6	....	....
7	7	....	....	7	7	....
8	8	8	....	7	8	....
9	8	9	....	8	9	....
10	9	10	....	9	10	10
11	10	11	11	9	10	11
12	10	11	12	10	11	12
13	11	12	13	10	12	13
14	12	13	14	11	12	13
15	12	13	14	12	13	14
16	13	14	15	12	14	15
17	13	15	16	13	14	16
18	14	15	17	13	15	16
19	15	16	17	14	15	17
20	15	17	18	15	16	18
21	16	17	19	15	17	18
22	17	18	19	16	17	19
23	17	19	20	16	18	20
24	18	19	21	17	19	20
25	18	20	21	18	19	21
26	19	20	22	18	20	22
27	20	21	23	19	20	22
28	20	22	23	19	21	23
29	21	22	24	20	22	24
30	21	23	25	20	22	24
31	22	24	25	21	23	25
32	23	24	26	22	24	26
33	23	25	27	22	24	26
34	24	25	27	23	25	27
35	24	26	28	23	25	27
36	25	27	29	24	26	28
37	25	27	29	24	27	29
38	26	28	30	25	27	29
39	27	28	31	26	28	30
40	27	29	31	26	28	31
41	28	30	32	27	29	31
42	28	30	32	27	29	32
43	29	31	33	28	30	32
44	29	31	34	28	31	33
45	30	32	34	29	31	34
46	31	33	35	30	32	34
47	31	33	36	30	32	35
48	32	34	36	31	33	36
49	32	34	37	31	34	36
50	33	35	37	32	34	37
60	39	41	44	37	40	43
70	44	47	50	43	46	49
80	50	52	56	48	51	55
90	55	58	61	54	57	61
100	61	64	67	59	63	66

STATISTICAL CHART 3

Variance ratio — 5 percent points for distribution of F

n_1 — degrees of freedom for numerator

n_2 — degrees of freedom for denominator

n_2 \ n_1	1	2	3	4	5	6	8	12	24	∞
1	161.4	199.5	215.7	224.6	230.2	234.0	238.9	243.9	249.0	254.3
2	18.51	19.00	19.16	19.25	19.30	19.33	19.37	19.41	19.45	19.50
3	10.13	9.55	9.28	9.12	9.01	8.94	8.84	8.74	8.64	8.53
4	7.71	6.94	6.59	6.39	6.26	6.16	6.04	5.91	5.77	5.63
5	6.61	5.79	5.41	5.19	5.05	4.95	4.82	4.68	4.53	4.36
6	5.99	5.14	4.76	4.53	4.39	4.28	4.15	4.00	3.84	3.67
7	5.59	4.74	4.35	4.12	3.97	3.87	3.73	3.57	3.41	3.23
8	5.32	4.46	4.07	3.84	3.69	3.58	3.44	3.28	3.12	2.93
9	5.12	4.26	3.86	3.63	3.48	3.37	3.23	3.07	2.90	2.71
10	4.96	4.10	3.71	3.48	3.33	3.22	3.07	2.91	2.74	2.54
11	4.84	3.98	3.59	3.36	3.20	3.09	2.95	2.79	2.61	2.40
12	4.75	3.88	3.49	3.26	3.11	3.00	2.85	2.69	2.50	2.30
13	4.67	3.80	3.41	3.18	3.02	2.92	2.77	2.60	2.42	2.21
14	4.60	3.74	3.34	3.11	2.96	2.85	2.70	2.53	2.35	2.13
15	4.54	3.68	3.29	3.06	2.90	2.79	2.64	2.48	2.29	2.07
16	4.49	3.63	3.24	3.01	2.85	2.74	2.59	2.42	2.24	2.01
17	4.45	3.59	3.20	2.96	2.81	2.70	2.55	2.38	2.19	1.96
18	4.41	3.55	3.16	2.93	2.77	2.66	2.51	2.34	2.15	1.92
19	4.38	3.52	3.13	2.90	2.74	2.63	2.48	2.31	2.11	1.88
20	4.35	3.49	3.10	2.87	2.71	2.60	2.45	2.28	2.08	1.84
21	4.32	3.47	3.07	2.84	2.68	2.57	2.42	2.25	2.05	1.81
22	4.30	3.44	3.05	2.82	2.66	2.55	2.40	2.23	2.03	1.78
23	4.28	3.42	3.03	2.80	2.64	2.53	2.38	2.20	2.00	1.76
24	4.26	3.40	3.01	2.78	2.62	2.51	2.36	2.18	1.98	1.73
25	4.24	3.38	2.99	2.76	2.60	2.49	2.34	2.16	1.96	1.71
26	4.22	3.37	2.98	2.74	2.59	2.47	2.32	2.15	1.95	1.69
27	4.21	3.35	2.96	2.73	2.57	2.46	2.30	2.13	1.93	1.67
28	4.20	3.34	2.95	2.71	2.56	2.44	2.29	2.12	1.91	1.65
29	4.18	3.33	2.93	2.70	2.54	2.43	2.28	2.10	1.90	1.64
30	4.17	3.32	2.92	2.69	2.53	2.42	2.27	2.09	1.89	1.62
40	4.08	3.23	2.84	2.61	2.45	2.34	2.18	2.00	1.79	1.51
60	4.00	3.15	2.76	2.52	2.37	2.25	2.10	1.92	1.70	1.39
120	3.92	3.07	2.68	2.45	2.29	2.17	2.02	1.83	1.61	1.25
∞	3.84	2.99	2.60	2.37	2.21	2.09	1.94	1.75	1.52	1.00

STATISTICAL CHART 3 — *Concluded*

Variance ratio — 1 percent point for distribution of *F*

n_1 — degrees of freedom for numerator

n_2 — degrees of freedom for denominator

n_2 \ n_1	1	2	3	4	5	6	8	12	24	∞
1	4052	4999	5403	5625	5764	5859	5981	6106	6234	6366
2	98.49	99.00	99.17	99.25	99.30	99.33	99.36	99.42	99.46	99.50
3	34.12	30.81	29.46	28.71	28.24	27.91	27.49	27.05	26.60	26.12
4	21.20	18.00	16.69	15.98	15.52	15.21	14.80	14.37	13.93	13.46
5	16.26	13.27	12.06	11.39	10.97	10.67	10.29	9.89	9.47	9.02
6	13.74	10.92	9.78	9.15	8.75	8.47	8.10	7.72	7.31	6.88
7	12.25	9.55	8.45	7.85	7.46	7.19	6.84	6.47	6.07	5.65
8	11.26	8.65	7.59	7.01	6.63	6.37	6.03	5.67	5.28	4.86
9	10.56	8.02	6.99	6.42	6.06	5.80	5.47	5.11	4.73	4.31
10	10.04	7.56	6.55	5.99	5.64	5.39	5.06	4.71	4.33	3.91
11	9.65	7.20	6.22	5.67	5.32	5.07	4.74	4.40	4.02	3.60
12	9.33	6.93	5.95	5.41	5.06	4.82	4.50	4.16	3.78	3.36
13	9.07	6.70	5.74	5.20	4.86	4.62	4.30	3.96	3.59	3.16
14	8.86	6.51	5.56	5.03	4.69	4.46	4.14	3.80	3.43	3.00
15	8.68	6.36	5.42	4.89	4.56	4.32	4.00	3.67	3.29	2.87
16	8.53	6.23	5.29	4.77	4.44	4.20	3.89	3.55	3.18	2.75
17	8.40	6.11	5.18	4.67	4.34	4.10	3.79	3.45	3.08	2.65
18	8.28	6.01	5.09	4.58	4.25	4.01	3.71	3.37	3.00	2.57
19	8.18	5.93	5.01	4.50	4.17	3.94	3.63	3.30	2.92	2.49
20	8.10	5.85	4.94	4.43	4.10	3.87	3.56	3.23	2.86	2.42
21	8.02	5.78	4.87	4.37	4.04	3.81	3.51	3.17	2.80	2.36
22	7.94	5.72	4.82	4.31	3.99	3.76	3.45	3.12	2.75	2.31
23	7.88	5.66	4.76	4.26	3.94	3.71	3.41	3.07	2.70	2.26
24	7.82	5.61	4.72	4.22	3.90	3.67	3.36	3.03	2.66	2.21
25	7.77	5.57	4.68	4.18	3.86	3.63	3.32	2.99	2.62	2.17
26	7.72	5.53	4.64	4.14	3.82	3.59	3.29	2.96	2.58	2.13
27	7.68	5.49	4.60	4.11	3.78	3.56	3.26	2.93	2.55	2.10
28	7.64	5.45	4.57	4.07	3.75	3.53	3.23	2.90	2.52	2.06
29	7.60	5.42	4.54	4.04	3.73	3.50	3.20	2.87	2.49	2.03
30	7.56	5.39	4.51	4.02	3.70	3.47	3.17	2.84	2.47	2.01
40	7.31	5.18	4.31	3.83	3.51	3.29	2.99	2.66	2.29	1.80
60	7.08	4.98	4.13	3.65	3.34	3.12	2.82	2.50	2.12	1.60
120	6.85	4.79	3.95	3.48	3.17	2.96	2.66	2.34	1.95	1.38
∞	6.64	4.60	3.78	3.32	3.02	2.80	2.51	2.18	1.79	1.00

STATISTICAL CHART 4

Significant studentized range at the 5% level

Degrees of freedom, *f*	Number of treatments, *a* 2	3	4	5	6	7	8	9	10	11	12	13	14	15	16	17	18	19	20
1	18.0	26.7	32.8	37.2	40.5	43.1	45.4	47.3	49.1	50.6	51.9	53.2	54.3	55.4	56.3	57.2	58.0	58.8	59.6
2	6.09	8.28	9.80	10.89	11.73	12.43	13.03	13.54	13.99	14.39	14.75	15.08	15.38	15.65	15.91	16.14	16.36	16.57	16.77
3	4.50	5.88	6.83	7.51	8.04	8.47	8.85	9.18	9.46	9.72	9.95	10.16	10.35	10.52	10.69	10.84	10.98	11.12	11.24
4	3.93	5.00	5.76	6.31	6.73	7.06	7.35	7.60	7.83	8.03	8.21	8.37	8.52	8.67	8.80	8.92	9.03	9.14	9.24
5	3.61	4.54	5.18	5.64	5.99	6.28	6.52	6.74	6.93	7.10	7.25	7.39	7.52	7.64	7.75	7.86	7.95	8.04	8.13
6	3.46	4.34	4.90	5.31	5.63	5.89	6.12	6.32	6.49	6.65	6.79	6.92	7.04	7.14	7.24	7.34	7.43	7.51	7.59
7	3.34	4.16	4.68	5.06	5.35	5.59	5.80	5.99	6.15	6.29	6.42	6.54	6.65	6.75	6.84	6.93	7.01	7.08	7.16
8	3.26	4.04	4.53	4.89	5.17	5.40	5.60	5.77	5.92	6.05	6.18	6.29	6.39	6.48	6.57	6.65	6.73	6.80	6.87
9	3.20	3.95	4.42	4.76	5.02	5.24	5.43	5.60	5.74	5.87	5.98	6.09	6.19	6.28	6.36	6.44	6.51	6.58	6.65
10	3.15	3.88	4.33	4.66	4.91	5.12	5.30	5.46	5.60	5.72	5.83	5.93	6.03	6.12	6.20	6.27	6.34	6.41	6.47
11	3.11	3.82	4.26	4.58	4.82	5.03	5.20	5.35	5.49	5.61	5.71	5.81	5.90	5.98	6.06	6.14	6.20	6.27	6.33
12	3.08	3.77	4.20	4.51	4.75	4.95	5.12	5.27	5.40	5.51	5.61	5.71	5.80	5.88	5.95	6.02	6.09	6.15	6.21
13	3.06	3.73	4.15	4.46	4.69	4.88	5.05	5.19	5.32	5.43	5.53	5.63	5.71	5.79	5.86	5.93	6.00	6.06	6.11
14	3.03	3.70	4.11	4.41	4.64	4.83	4.99	5.13	5.25	5.36	5.46	5.56	5.64	5.72	5.79	5.86	5.92	5.98	6.03
15	3.01	3.67	4.08	4.37	4.59	4.78	4.94	5.08	5.20	5.31	5.40	5.49	5.57	5.65	5.72	5.79	5.85	5.91	5.96
16	3.00	3.65	4.05	4.34	4.56	4.74	4.90	5.03	5.15	5.26	5.35	5.44	5.52	5.59	5.66	5.73	5.79	5.84	5.90
17	2.98	3.62	4.02	4.31	4.52	4.70	4.86	4.99	5.11	5.21	5.31	5.39	5.47	5.55	5.61	5.68	5.74	5.79	5.84
18	2.97	3.61	4.00	4.28	4.49	4.67	4.83	4.96	5.07	5.17	5.27	5.35	5.43	5.50	5.57	5.63	5.69	5.74	5.79
19	2.96	3.59	3.98	4.26	4.47	4.64	4.79	4.92	5.04	5.14	5.23	5.32	5.39	5.46	5.53	5.59	5.65	5.70	5.75
20	2.95	3.58	3.96	4.24	4.45	4.62	4.77	4.90	5.01	5.11	5.20	5.28	5.36	5.43	5.50	5.56	5.61	5.66	5.71
24	2.92	3.53	3.90	4.17	4.37	4.54	4.68	4.81	4.92	5.01	5.10	5.18	5.25	5.32	5.38	5.44	5.50	5.55	5.59
30	2.89	3.48	3.84	4.11	4.30	4.46	4.60	4.72	4.83	4.92	5.00	5.08	5.15	5.21	5.27	5.33	5.38	5.43	5.48
40	2.86	3.44	3.79	4.04	4.23	4.39	4.52	4.63	4.74	4.82	4.90	4.98	5.05	5.11	5.17	5.22	5.27	5.32	5.36
60	2.83	3.40	3.74	3.98	4.16	4.31	4.44	4.55	4.65	4.73	4.81	4.88	4.94	5.00	5.06	5.11	5.15	5.20	5.24
120	2.80	3.36	3.69	3.92	4.10	4.24	4.36	4.47	4.56	4.64	4.71	4.78	4.84	4.90	4.95	5.00	5.04	5.09	5.13
∞	2.77	3.32	3.63	3.86	4.03	4.17	4.29	4.39	4.47	4.55	4.62	4.68	4.74	4.80	4.84	4.89	4.93	4.97	5.01

STATISTICAL CHART 5

Rank totals

Rank totals required for significance at the 5% level ($P \leqslant 0.05$). The four figure blocks sent: lowest insignificant rank sum, any treatment–highest insignificant rank sum, any ment. Lowest insignificant rank sum, predetermined treatment–highest insignificant rank predetermined treatment.

No. of reps.	N							
	2	3	4	5	6	7	8	9
2	....	....	....	 3-9	 3-11	 3-13	 4-14	 4-16
3	....	 4-8	 4-11	4-14 5-13	4-17 6-15	4-20 6-18	4-23 7-20	5-25 8-22
4	....	5-11 5-11	5-15 6-14	6-18 7-17	6-22 8-20	7-25 9-23	7-29 10-26	8-32 11-29
5	 6-9	6-14 7-13	7-18 8-17	8-22 10-20	9-26 11-24	9-31 13-27	10-35 14-31	11-39 15-35
6	7-11 7-11	8-16 9-15	9-21 11-19	10-26 12-24	11-31 14-28	12-36 16-32	13-41 18-36	14-46 20-40
7	8-13 8-13	10-18 10-18	11-24 13-22	12-30 15-27	14-35 17-32	15-41 19-37	17-46 22-41	18-52 24-46
8	9-15 10-14	11-21 12-20	13-27 15-25	15-33 17-31	17-39 20-36	18-46 23-41	20-52 25-47	22-58 28-52
9	11-16 11-16	13-23 14-22	15-30 17-28	17-37 20-34	19-44 23-40	22-50 26-46	24-57 29-52	26-64 32-58
10	12-18 12-18	15-25 16-24	17-33 19-31	20-40 23-37	22-48 26-44	25-55 30-50	27-63 33-57	30-70 37-63
11	13-20 14-19	16-28 18-26	19-36 21-34	22-44 25-41	25-52 29-48	28-60 33-55	31-68 37-62	34-76 41-69
12	15-21 15-21	18-30 19-29	21-39 24-36	25-47 28-44	28-56 32-52	31-65 37-59	34-74 41-67	38-82 45-75
13	16-23 17-22	20-32 21-31	24-41 26-39	27-51 31-47	31-60 35-56	35-69 40-64	38-79 45-72	42-88 50-80
14	17-25 18-24	22-34 23-33	26-44 28-42	30-54 33-51	34-64 38-60	38-74 44-68	42-84 49-77	46-94 54-86
15	19-26 19-26	23-37 25-35	28-47 30-45	32-58 36-54	37-68 42-63	41-79 47-73	46-89 53-82	50-100 59-91
16	20-28 21-27	25-39 27-37	30-50 33-47	35-61 39-57	40-72 45-67	45-83 51-77	49-95 57-87	54-106 63-97
17	22-29 22-29	27-41 28-40	32-53 35-50	38-64 41-61	43-76 48-71	48-88 54-82	53-100 61-92	58-112 67-103
18	23-31 24-30	29-43 30-42	34-56 37-53	40-68 44-64	46-80 51-75	51-93 58-86	57-105 65-97	62-118 72-108
19	24-33 25-32	30-46 32-44	37-58 39-56	43-71 47-67	49-84 54-79	55-97 62-90	61-110 69-102	67-123 76-114

s, or samples ranked

12	13	14	15	16	17	18	19	20
....	....	....	....	....	....	....	....	3-39
5-21	5-23	5-25	6-26	6-28	6-30	7-31	7-33	7-35
5-34	5-37	5-40	6-42	6-45	6-48	6-51	6-54	7-56
10-29	10-32	11-34	12-36	12-39	13-41	14-43	14-46	15-48
9-43	9-47	10-50	10-54	10-58	11-61	11-65	12-68	12-72
15-37	16-40	17-43	18-46	19-49	20-52	21-55	22-58	23-61
13-52	14-56	14-61	15-65	16-69	16-74	17-78	18-82	18-87
20-45	21-49	23-52	24-56	25-60	27-63	28-67	30-70	31-74
18-60	19-65	19-71	20-76	21-81	22-86	23-91	24-96	25-101
25-53	27-57	29-61	31-65	32-70	34-74	36-78	38-82	40-86
22-69	23-75	25-80	26-86	27-92	29-97	30-103	31-109	32-115
30-61	33-65	35-70	37-75	39-80	42-84	44-89	46-94	48-99
27-77	29-83	30-90	32-96	33-103	35-109	37-115	38-122	40-128
36-68	39-73	41-79	44-84	47-89	49-95	52-100	54-106	57-111
32-85	34-92	36-99	38-106	40-113	42-120	44-127	45-135	47-142
41-76	45-81	48-87	51-93	54-99	57-105	60-111	63-117	66-123
37-93	39-101	41-109	44-116	46-124	48-132	51-139	53-147	55-155
47-83	51-89	54-96	57-103	61-109	64-116	68-122	71-129	75-135
42-101	45-109	47-118	50-126	53-134	55-143	58-151	60-160	63-168
53-90	57-97	60-105	64-112	68-119	72-126	76-133	80-140	84-147
47-109	50-118	53-127	56-136	59-145	62-154	65-163	68-172	71-181
58-98	63-105	67-113	71-121	76-128	80-136	84-144	89-151	93-159
52-117	56-126	59-136	62-146	66-155	69-165	73-174	76-184	79-194
64-105	69-113	74-121	78-130	83-138	88-146	93-154	97-163	102-171
57-125	61-135	65-145	69-155	73-165	76-176	80-186	84-196	88-206
70-112	75-121	80-130	85-139	91-147	96-156	101-165	106-174	111-183
63-132	67-143	71-154	75-165	79-176	84-186	88-197	92-208	96-219
75-120	81-129	87-138	92-148	98-157	104-166	109-176	115-185	121-194
68-140	73-151	77-163	82-174	86-186	91-197	95-209	100-220	104-232
81-127	87-137	93-147	100-156	106-166	112-176	118-186	124-196	130-206
73-148	78-160	83-172	88-184	93-196	98-208	103-220	108-232	113-244
87-134	94-144	100-155	107-165	113-176	120-186	126-197	133-207	139-218
79-155	84-168	90-180	95-193	100-206	106-218	111-231	116-244	121-257
93-141	100-152	107-163	114-174	121-185	128-196	135-207	142-218	149-229
84-163	90-176	96-189	102-202	107-216	113-229	119-242	124-256	130-269
99-148	106-160	114-171	121-183	128-195	136-206	143-218	151-229	158-241

STATISTICAL CHART 6

Scores for ranked data

The mean deviations of the 1st, 2nd, 3rd . . . largest members of samples of different sizes; zero and negative values omitted.

Ordinal number	Size of Sample								
	—	2	3	4	5	6	7	8	9	10
1		0.56	0.85	1.03	1.16	1.27	1.35	1.42	1.49	1.54
2				0.30	0.50	0.64	0.76	0.85	0.93	1.00
3						0.20	0.35	0.47	0.57	0.66
4								0.15	0.27	0.38
5										0.12

	11	12	13	14	15	16	17	18	19	20
1	1.59	1.63	1.67	1.70	1.74	1.76	1.79	1.82	1.84	1.87
2	1.06	1.12	1.16	1.21	1.25	1.28	1.32	1.35	1.38	1.41
3	0.73	0.79	0.85	0.90	0.95	0.99	1.03	1.07	1.10	1.13
4	0.46	0.54	0.60	0.66	0.71	0.76	0.81	0.85	0.89	0.92
5	0.22	0.31	0.39	0.46	0.52	0.57	0.62	0.67	0.71	0.75
6		0.10	0.19	0.27	0.34	0.39	0.45	0.50	0.55	0.59
7				0.09	0.17	0.23	0.30	0.35	0.40	0.45
8						0.08	0.15	0.21	0.26	0.31
9								0.07	0.13	0.19
10										0.06

	21	22	23	24	25	26	27	28	29	30
1	1.89	1.91	1.93	1.95	1.97	1.98	2.00	2.01	2.03	2.04
2	1.43	1.46	1.48	1.50	1.52	1.54	1.56	1.58	1.60	1.62
3	1.16	1.19	1.21	1.24	1.26	1.29	1.31	1.33	1.35	1.36
4	0.95	0.98	1.01	1.04	1.07	1.09	1.11	1.14	1.16	1.18
5	0.78	0.82	0.85	0.88	0.91	0.93	0.96	0.98	1.00	1.03
6	0.63	0.67	0.70	0.73	0.76	0.79	0.82	0.85	0.87	0.89
7	0.49	0.53	0.57	0.60	0.64	0.67	0.70	0.73	0.75	0.78
8	0.36	0.41	0.45	0.48	0.52	0.55	0.58	0.61	0.64	0.67
9	0.24	0.29	0.33	0.37	0.41	0.44	0.48	0.51	0.54	0.57
10	0.12	0.17	0.22	0.26	0.30	0.34	0.38	0.41	0.44	0.47
11		0.06	0.11	0.16	0.20	0.24	0.28	0.32	0.35	0.38
12				0.05	0.10	0.14	0.19	0.22	0.26	0.29
13						0.05	0.09	0.13	0.17	0.21
14								0.04	0.09	0.12
15										0.04

Tests of psychological preference and some other experimental data suffice to place a series of magnitudes in order of preference, without supplying metrical values. Analyses of variance, correlations, etc., can be carried out on such data by using the normal scores, appropriate to each position in order, in a sample of the size observed. Ties may be scored with the means of the ordinal values involved, but in such cases the sums of squares given will require correction.

STATISTICAL CHART 7

The distribution of *t*

Degrees of freedom	Probability of a larger value, sign ignored								
	0.500	0.400	0.200	0.100	0.050	0.025	0.010	0.005	0.001
1	1.000	1.376	3.078	6.314	12.706	25.452	63.657		
2	0.816	1.061	1.886	2.920	4.303	6.205	9.925	14.089	31.598
3	0.765	0.978	1.638	2.353	3.182	4.176	5.841	7.453	12.941
4	0.741	0.941	1.533	2.132	2.776	3.495	4.604	5.598	8.610
5	0.727	0.920	1.476	2.015	2.571	3.163	4.032	4.773	6.859
6	0.718	0.906	1.440	1.943	2.447	2.969	3.707	4.317	5.959
7	0.711	0.896	1.415	1.895	2.365	2.841	3.499	4.029	5.405
8	0.706	0.889	1.397	1.860	2.306	2.752	3.355	3.832	5.041
9	0.703	0.883	1.383	1.833	2.262	2.685	3.250	3.690	4.781
10	0.700	0.879	1.372	1.812	2.228	2.634	3.169	3.581	4.587
11	0.697	0.876	1.363	1.796	2.201	2.593	3.106	3.497	4.437
12	0.695	0.873	1.356	1.782	2.179	2.560	3.055	3.428	4.318
13	0.694	0.870	1.350	1.771	2.160	2.533	3.012	3.372	4.221
14	0.692	0.868	1.345	1.761	2.145	2.510	2.977	3.326	4.140
15	0.691	0.866	1.341	1.753	2.131	2.490	2.947	3.286	4.073
16	0.690	0.865	1.337	1.746	2.120	2.473	2.921	3.252	4.015
17	0.689	0.863	1.333	1.740	2.110	2.458	2.898	3.222	3.965
18	0.688	0.862	1.330	1.734	2.101	2.445	2.878	3.197	3.922
19	0.688	0.861	1.328	1.729	2.093	2.433	2.861	3.174	3.883
20	0.687	0.860	1.325	1.725	2.086	2.423	2.845	3.153	3.850
21	0.686	0.859	1.323	1.721	2.080	2.414	2.831	3.135	3.819
22	0.686	0.858	1.321	1.717	2.074	2.406	2.819	3.119	3.792
23	0.685	0.858	1.319	1.714	2.069	2.398	2.807	3.104	3.767
24	0.685	0.857	1.318	1.711	2.064	2.391	2.797	3.090	3.745
25	0.684	0.856	1.316	1.708	2.060	2.385	2.787	3.078	3.725
26	0.684	0.856	1.315	1.706	2.056	2.379	2.779	3.067	3.707
27	0.684	0.855	1.314	1.703	2.052	2.373	2.771	3.056	3.690
28	0.683	0.855	1.313	1.701	2.048	2.368	2.763	3.047	3.674
29	0.683	0.854	1.311	1.699	2.045	2.364	2.756	3.038	3.659
30	0.683	0.854	1.310	1.697	2.042	2.360	2.750	3.030	3.646
35	0.682	0.852	1.306	1.690	2.030	2.342	2.724	2.996	3.591
40	0.681	0.851	1.303	1.684	2.021	2.329	2.704	2.971	3.551
45	0.680	0.850	1.301	1.680	2.014	2.319	2.690	2.952	3.520
50	0.680	0.849	1.299	1.676	2.008	2.310	2.678	2.937	3.496
55	0.679	0.849	1.297	1.673	2.004	2.304	2.669	2.925	3.476
60	0.679	0.848	1.296	1.671	2.000	2.299	2.660	2.915	3.460
70	0.678	0.847	1.294	1.667	1.994	2.290	2.648	2.899	3.435
80	0.678	0.847	1.293	1.665	1.989	2.284	2.638	2.887	3.416
90	0.678	0.846	1.291	1.662	1.986	2.279	2.631	2.878	3.402
100	0.677	0.846	1.290	1.661	1.982	2.276	2.625	2.871	3.390
120	0.677	0.845	1.289	1.658	1.980	2.270	2.617	2.860	3.373
∞	0.6745	0.8416	1.2816	1.6448	1.9600	2.2414	2.5758	2.8070	3.2905

STATISTICAL CHART 8

Table of random numbers — permutations of nine

98119	47634	62128	74824	26316	69967	99242
42293	62781	39637	56945	93661	35153	26837
71926	19563	58873	41611	12194	24228	17798
17455	58857	11764	19452	57975	47815	52523
66834	25245	27285	25299	71782	88679	34114
29662	83196	93516	32777	64843	92581	73375
35341	74918	44949	93188	85428	71742	68486
84787	96479	76491	68563	38259	16396	81659
53578	31322	85352	87336	49537	53434	45961
24814	99952	56378	32381	21148	97297	72848
52498	87383	22131	69919	15466	36866	98794
71675	78536	73495	27778	77622	55159	51263
98581	33164	49769	86257	88255	72928	85426
37137	45775	97913	51435	93913	14312	16975
63226	16691	38586	93122	54874	29581	44112
86349	64849	61622	15644	62331	63645	39681
19752	22217	85847	48893	46597	48774	63557
45963	51428	14254	74566	39789	81433	27339
35951	41886	65455	39863	86891	26531	12772
62737	28622	97774	94399	65615	71385	36396
93289	53491	21397	16212	98532	12463	83527
57693	75743	49661	77747	54358	44124	59213
74862	87515	54246	43585	19779	87859	64959
19174	39259	33113	82151	47924	59642	75834
86518	66168	12838	25974	31147	93998	41141
48426	14937	88522	68628	73483	38217	97468
21345	92374	76989	51436	22266	65776	28685
46622	22565	57564	62716	48346	22575	76356
32234	77979	61291	19143	19222	98313	54127
25553	66127	89656	47931	84577	46781	43718
14871	33813	48879	33229	37769	79698	38599
91485	98384	23448	75582	95118	67862	82865
57997	15656	36185	56864	21884	53946	61484
78366	84292	74722	28657	76693	84137	99632
63148	49438	15937	81498	52455	31229	17273
89719	51741	92313	94375	63931	15454	25941
85392	17996	58885	38247	84138	71165	44722
72575	99477	91117	93856	77347	82872	29147
51457	72341	72394	47919	62519	34731	82898
96724	46815	23931	75785	95794	15923	57213
48683	28624	46578	52168	11983	99488	61586
64968	51183	64763	19332	33622	27299	73355
27846	64569	85256	81471	49461	58617	95634
19211	35232	19449	26624	58256	66356	18461
33139	83758	37622	64593	26875	43544	36979

STATISTICAL CHART 8 — *Concluded*

Random numbers — permutations of nine

42659	14978	74643	21224	33681	47164	99323
68131	96442	21839	59659	22718	79895	24254
36478	75184	92278	74478	19924	63749	61566
14824	37556	35982	63737	45539	56252	46132
79966	61713	57561	85393	54495	38978	73771
95313	43327	16415	42881	97242	84586	58488
21585	82635	43154	16545	86363	91337	82695
53797	28899	69727	38112	61157	12611	37847
87242	59261	88396	97966	78876	25423	15919
16383	72679	78165	11448	12781	89769	75817
22898	25526	34851	48721	65122	95142	39438
87751	97953	43779	55296	27956	53894	62683
35614	36891	15414	87175	88867	72978	84774
93275	41737	91937	24934	79478	36256	17991
44946	18212	22682	79363	46615	24425	26256
78167	54368	69326	36587	34349	68617	91365
61439	89445	56293	92612	91594	41581	58129
59522	63184	87548	63859	53233	17333	43542
49661	11831	37549	97499	94883	32513	95688
53196	68259	65492	28563	21942	86426	81796
86857	75113	73927	69736	86511	95998	73811
37442	22945	91338	12117	39629	48254	12377
98339	49672	86783	81928	17356	53331	29969
24714	33386	29114	36371	42134	69875	54255
15928	96568	14661	55252	75467	11189	47432
71583	54794	58875	74885	53795	27767	38544
62275	87427	42256	43644	68278	74642	66123
31727	54363	98644	86696	58126	54111	12173
22588	96555	31488	39317	73757	67449	37334
15869	22124	49991	13468	84674	28392	89592
63276	85881	75722	45251	12565	72976	44247
98414	73738	64539	57729	36299	46527	76481
57633	41279	52277	94144	21331	19263	23856
46155	17446	13115	68983	67448	33855	98668
84991	39692	86853	21575	45912	85738	51719
79342	68917	27366	72832	99883	91684	65925
92495	48448	19485	27965	98734	38213	35326
11813	86599	27677	68698	22229	14862	28984
39557	24933	81923	76577	67867	25957	14118
86672	69156	96531	11751	83458	93428	51672
75339	73687	68254	34146	59592	62575	69737
24921	97224	42748	83432	46971	77696	72261
43784	51812	73399	99219	31113	89734	43549
68166	35771	55116	52383	15686	46389	86495
57248	12365	34862	45824	74345	51141	97853

CONVERSION FACTORS FOR METRIC SYSTEM

Imperial units	Approximate conversion factor	Results in:	
LINEAR			
inch	x 25	**millimetre**	**(mm)**
foot	x 30	**centimetre**	**(cm)**
yard	x 0.9	**metre**	**(m)**
mile	x 1.6	**kilometre**	**(km)**
AREA			
square inch	x 6.5	**square centimetre**	**(cm^2)**
square foot	x 0.09	**square metre**	**(m^2)**
acre	x 0.40	**hectare**	**(ha)**
VOLUME			
cubic inch	x 16	**cubic centimetre**	**(cm^3)**
cubic foot	x 28	**cubic decimetre**	**(dm^3)**
cubic yard	x 0.8	**cubic metre**	**(m^3)**
fluid ounce	x 28	**millilitre**	**(mL)**
pint	x 0.57	**litre**	**(L)**
quart	x 1.1	**litre**	**(L)**
gallon	x 4.5	**litre**	**(L)**
WEIGHT			
ounce	x 28	**gram**	**(g)**
pound	x 0.45	**kilogram**	**(kg)**
short ton (2000 lb)	x 0.9	**tonne**	**(t)**
TEMPERATURE			
degrees Fahrenheit	(°F-32) x 0.56 or (°F-32) x 5/9	**degrees Celsius**	**(°C)**
PRESSURE			
pounds per square inch	x 6.9	**kilopascal**	**(kPa)**
POWER			
horsepower	x 746	**watt**	**(W)**
	x 0.75	**kilowatt**	**(kW)**
SPEED			
feet per second	x 0.30	**metres per second**	**(m/s)**
miles per hour	x 1.6	**kilometres per hour**	**(km/h)**
AGRICULTURE			
gallons per acre	x 11.23	**litres per hectare**	**(L/ha)**
quarts per acre	x 2.8	**litres per hectare**	**(L/ha)**
pints per acre	x 1.4	**litres per hectare**	**(L/ha)**
fluid ounces per acre	x 70	**millilitres per hectare**	**(mL/ha)**
tons per acre	x 2.24	**tonnes per hectare**	**(t/ha)**
pounds per acre	x 1.12	**kilograms per hectare**	**(kg/ha)**
ounces per acre	x 70	**grams per hectare**	**(g/ha)**
plants per acre	x 2.47	**plants per hectare**	**(plants/ha)**